THE GENE

THE GENE

A Historical Perspective

By Ted Everson

Greenwood Guides to Great Ideas in Science
Brian Baigrie, Series Editor

Greenwood Press
Westport, Connecticut • London

Library of Congress Cataloging-in-Publication Data

Everson, Ted.
    The gene : a historical perspective / by Ted Everson.
        p. cm. — (Greenwood guides to great ideas in science, ISSN: 1559–5374)
    Includes bibliographical references.
    ISBN-13 : 978–0–313–33449–8 (alk. paper)
    ISBN-10 : 0–313–33449–8 (alk. paper)
    1. Genetics—History. 2. Molecular genetics—History. 3. Heredity.
I. Title.
    QH428.E84   2007
    576.509—dc22   2007003661

British Library Cataloguing in Publication Data is available.

Library of Congress Catalog Card Number: 2007003661
ISBN-13: 978–0–313–33449–8
ISBN-10: 0–313–33449–8
ISSN: 1559–5374

First published in 2007

Greenwood Press, 88 Post Road West, Westport, CT 06881
An imprint of Greenwood Publishing Group, Inc.
www.greenwood.com

Printed in the United States of America

CONTENTS

LIST OF ILLUSTRATIONS

SERIES FOREWORD

The volumes in this series are devoted to concepts that are fundamental to different branches of the natural sciences—the gene, the quantum, geological cycles, planetary motion, evolution, the cosmos, and forces in nature, to name just a few. Although these volumes focus on the historical development of scientific ideas, the underlying hope of this series is that the reader will gain a deeper understanding of the process and spirit of scientific practice. In particular, in an age in which students and the public have been caught up in debates about controversial scientific ideas, it is hoped that readers of these volumes will better appreciate the provisional character of scientific truths by discovering the manner in which these truths were established.

   The history of science as a distinctive field of inquiry can be traced to the early seventeenth century when scientists began to compose histories of their own fields. As early as 1601, the astronomer and mathematician Johannes Kepler composed a rich account of the use of hypotheses in astronomy. During the ensuing three centuries, these histories were increasingly integrated into elementary textbooks, the chief purpose of which was to pinpoint the dates of discoveries as a way of stamping out all too frequent propriety disputes and to highlight the errors of predecessors and contemporaries. Indeed, historical introductions in scientific textbooks continued to be common well into the twentieth century. Scientists also increasingly wrote histories of their disciplines—separate from those that appeared in textbooks—to explain to a broad popular audience the basic concepts of their science.

   The history of science remained under the auspices of scientists until the establishment of the field as a distinct professional activity in the middle of the twentieth century. As academic historians assumed control of history of science writing, they expended enormous energies in the attempt to forge a distinct and autonomous discipline. The result of this struggle to position the history of science as an intellectual endeavor that was valuable in its own right, and not merely in consequence of its ties to science, was that historical studies of the natural sciences were no longer composed with an eye toward educating a wide audience that included nonscientists but instead were composed with the aim of being consumed by other professional historians of science. And as historical breadth was sacrificed for technical detail, the literature became increasingly daunting in its technical detail. While this scholarly work increased our understanding of the nature of science, the technical demands imposed on the reader had the unfortunate consequence of leaving behind the general reader.

As Series Editor, my ambition for these volumes is that they will combine the best of these two types of writing about the history of science. In step with the general introductions that we associate with historical writing by scientists, the purpose of these volumes is educational—they have been authored with the aim of making these concepts accessible to students—high school, college, and university—and to the general public. However, the scholars who have written these volumes are not only able to impart genuine enthusiasm for the science discussed in the volumes of this series, they can use the research and analytic skills that are the staples of any professional historian and philosopher of science to trace the development of these fundamental concepts. My hope is that a reader of these volumes will share some of the excitement of these scholars—for both science and its history.

Brian Baigrie
University of Toronto
Series Editor

PREFACE

The "gene" as a unit of hereditary information was first conceptualized in the early twentieth century; by midcentury the gene was discovered to be physically located within a molecule called deoxyribonucleic acid, or DNA; by the end of the century, DNA had become an object of experimentation and manipulation, as genetic engineering promises to revolutionize biology and medicine. The twentieth century was, indeed, the century of the gene. However, the concept of heredity has a much deeper history, spanning the entire history of Western thought. All of the great ages of Western civilization produced thinkers who struggled to understand why offspring look like their parents and generally why living things display the remarkable ability to faithfully reproduce themselves over countless generations. Was there something special about life that allowed for such a feat? These issues have been central to several millenia of speculation, observation, and experimentation regarding the nature of heredity; this book explores some of these activities, in the process revealing remarkably consistent issues that people have struggled with in order to understand inheritance.

I gratefully acknowledge Brian Baigrie, Series Editor for the Greenwood Guide to great Ideas in the History of Science, and Kevin Downing, Senior Acquisitions Editor at Greenwood Publishing, for their invaluable assistance and encouragement.

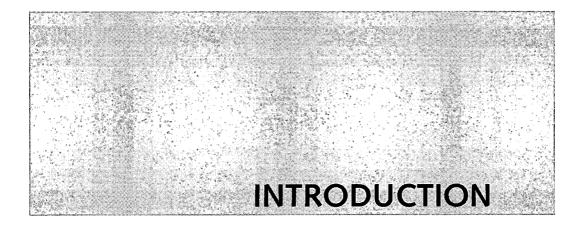

INTRODUCTION

Few scientific concepts today have captured the interest and attention of biologists, historians and philosophers of science, policymakers, the business community, and the general public as much as that of the gene. Today and historically, people interpret the gene in many different ways: from its early depictions as simply a by-product or outgrowth of an organism's development, to its descriptions in the early twentieth century as a physical unit of hereditary information, to today's primary conception (and there is in fact more than one) as a complex assemblage of subunits of deoxyribonucleic acid (DNA), the gene's identity has changed and evolved over time. And these changes, as we shall see, often reflected other underlying views of how the world worked—or how people thought it *should* work.

The concept of the gene thus has a historical lineage, determined by a broad array of scientific, technological, and social events that span centuries. Our knowledge of this history also has a historical lineage, and historians have become increasingly aware of the complexities of interaction of historical events that have gone into forming our modern concept of the gene. This historical perspective has taught us many things: for one, we can expect that, in the future as in the past, our conception of the gene will likely change in currently unpredictable ways. Even in the past 20 to 30 years, we have seen significant changes in scientists' views about what a gene is.

The study of the gene as a specific branch of science—called *genetics*—is a very recent phenomenon, beginning in the early twentieth century, when several scientists realized that the late-nineteenth-century breeding experiments of an obscure monk named Gregor Mendel shed light on their questions about heredity (see Chapter 4). The term *genetics* was coined shortly later. Despite its recent coinage, however, genetics has deep historical roots. Like most branches of modern science, historians have looked to the ancient Greeks for some of the origins of genetics; Chapter 1 describes these origins. Conceptions of heredity were at the forefront of ancient Greek thought, intertwined with debates between those who saw in life some uniquely mysterious power to predictably grow and reproduce and those who sought to explain away these qualities as the visible manifestation of random matter in motion. This conceptual divide in understanding life remained at the forefront of much of natural philosophy for two thousand years, until our modern conceptions of genetics and evolution took root in the early twentieth century. This is not to say that concepts of life, and of heredity, were static for two thousand years. For example, Chapter 1 also describes some of the ways in which the Middle Ages witnessed a great variety of interpretation and modification of ancient Greek thought. Once thought of as the "Dark Ages," today the medieval period is understood as one of rich intellectual activity in its own right, although the early medieval centers of activity were not in Western Europe but in lands once belonging to the vast Islamic Empire. Later Western Europe became exposed to this intellectual legacy, using and modifying it just as Muslim scholars used ancient Greek and Roman works.

The development of a Western intellectual tradition saw a rapid quickening of pace during the Renaissance and the Enlightenment, the subject of Chapter 2. In the context of remarkable social, political, and economic change, Western Europe saw the formation and institutionalization of a distinct field of human endeavor—*science*—with specialized practitioners and distinct social organizations. It was during the Enlightenment that the term *science* came to be associated with a specific and relatively narrow sort of activity, with which we associate the term today. A major component of this activity was the study of life, and the term *biology* began to be used during the Enlightenment as well, to refer to the study of living things. Some of the most important names in the history of biology, familiar to anyone who has taken an introductory biology course, come from this time period.

By the beginning of the nineteenth century—the subject of Chapter 3 —science was an established and highly respected practice, having full cultural authority over the practice of knowledge production. The nineteenth century saw work of monumental importance in the history of biology, which today forms the foundation of modern biology, including Charles Darwin's theory of evolution by natural selection, Gregor Mendel's research that later illustrated the existence of genes, and the discovery of chromosomes by August Weismann and others. Yet a great deal of confusion has characterized the popular understanding of these men. Chapters 3 and 4 describe these events and their extraordinary effect on the subsequent developments in our understanding of heredity. Chapter 4, in particular, focuses on the extraordinary story of

Gregor Mendel—his goals as a breeder of pea plants; his unexpected results and tentative conclusions; the "rediscovery" in the early twentieth century of Mendel's findings—and we shall see, this word does not quite capture the nuances of history—and, finally, lessons learned from the Mendel story for our own understandings of history, science, and the history of science.

Which brings us to the twentieth century and the birth of modern genetics. Any book devoted to the concept of the gene must focus primarily on the twentieth century, for it is from the past hundred years or so that the bulk of modern genetics has attained its guiding principles and concepts and its social and institutional foundations. In the twentieth century, biologists focused ever more on the physical reality of the gene, and at the same time biological science grew into a central institution of modern society, gaining a great deal of influence in the understanding of how life—including human life—works. And, as we shall see, in the early twentieth century genetics became a tool for several groups to argue for their views about how to improve society. The practice of *eugenics*, which involved attempts by those with power to control the reproductive patterns of those they assumed were genetically inferior, is long discredited but remains an important component of any understanding of the history of the gene.

As Chapter 5 illustrates, in the 1930s the gene changed from simply a concept to something with a physical form when Thomas Hunt Morgan and colleagues discovered that heredity seemed to be controlled by very specific, discrete molecular units, located on the newly discovered chromosomes. Morgan's studies catalyzed the search for the physical gene, and methods for the analysis of chromosomes were some of the earliest developments in the new science of "molecular biology," coined in the late 1930s as a new research program that would aggressively seek to understand the underlying molecular mechanisms that governed life's processes. Key to this research program was an attempt to discover the molecular basis of inheritance. Molecular biology, described in Chapter 6, came to dominate biological research, in part because of a strong interest by funders of the field to get at the heart of what they thought were crucial questions of social importance: what explains the many social problems that plagued humanity? In the wake of the eugenics movement, which was falling out of favor as a scientific practice by the 1930s, there remained an interest in understanding the biology of social development; molecular biology was born under what was called the "Science of Man" proposal of the Rockefeller Foundation, a major funder of the sciences, in order to address social issues using principles of the new biology. How compelling is this argument that biology is an important tool in understanding our social problems? And why has this idea been so resilient, remaining with us even today? Chapter 6 addresses this important social issue.

Because Morgan had determined that genes were located on chromosomes, molecular biology initially focused on the study of proteins, the major component of chromosomes. The other component was DNA, a simple and chemically inert molecule that seemed much less interesting than the diverse, complex, and highly active proteins. But, in the 1940s, a few molecular biologists began to focus on DNA, suggesting that it, rather than protein, might be the physical source of the elusive gene. Chapter 7 describes this research, culminating in 1953 with the publication of James Watson and Francis Crick's model of the double helical structure of DNA, one of the most famous discoveries of the history of science. Few stories in the annals of science have been repeated more than this discovery, which has become the foundation on which modern genetics and biotechnology rest. But in recent years there have emerged new understandings of this story. In particular, the work of Rosalind Franklin, a woman studying DNA structure at King's College in London, was crucial to Watson and Crick's work. These new understandings of Franklin's importance provide valuable lessons to us about how science works and illustrate the many human dimensions underlying major scientific discoveries.

Following the discovery of DNA's structure, a great deal of activity centered on how it worked. Where was the gene in this molecule? How was DNA related to the mechanism of heredity? Chapter 8 describes the developments in the 1960s that culminated in an understanding of how DNA managed to faithfully replicate itself over countless generations and how it encoded all of the information required to make a new life form. This was an exhilarating time in the history of gene research, as molecular biologists became progressively more excited and awed by the potential capacity to understand and experiment with DNA. Discoveries were made at a dizzying pace, culminating in the landmark achievements in manipulating DNA, described in Chapter 9. Genetic engineering, gene splicing, recombinant DNA technology—all of these terms and more have been used to describe these momentous developments. A major new industry, *biotechnology*, grew out of these achievements.

Never before had it been possible to mix DNA from vastly different species of life. Today, it is a relatively trivial task to attach human genes to, for example, bacterial or yeast genes and to grow an organism with the mixed genes functioning properly. Connected with these first genetic engineering experiments were a variety of scientific, social, political and economic developments, as scientists and the broader public grappled with the implications inherent in this new power to manipulate what was seen by many, scientists and nonscientists alike, as the essence of life itself.

Close on the heels of the development of biotechnology came one of the most ambitious developments of modern science: the Human Genome Project, a large-scale, multinational effort to discover all of the genes, and sequence all of the DNA, contained in the cells of human beings (and other animals).

The Human Genome Project, the subject of Chapter 10, is a fascinating example of the complexities inherent in the history of science. Science and politics were heavily intertwined in the history of this ambitious initiative. The story is also a remarkable illustration of how the concept of the gene has evolved over time. Discoveries connected to the Human Genome Project reveal that DNA, and the vast cellular machinery that is related to its function, is a remarkably complex system that cannot be reduced to any assumption that the gene is an all-powerful "master control" that can explain life itself.

Yet, at the same time, modern biology and medicine have increasingly come to be dominated by genetics. Health and illness, for example, have increasingly become understood in terms of the concept of the gene, as increasingly powerful analytic techniques suggest that many people have a "genetic susceptibility" to one or other of the most common illnesses affecting humanity or that experimentation and intervention at a genetic level can resolve previously insoluble health issues in society. The concept of the gene is thus more powerful today than ever, even while the complexity of its biology is increasingly understood. Chapter 11 discusses a variety of topics that are part of modern interest in the gene in the wake of the Human Genome Project and considers some important modern debates about the centrality of the gene to many areas of human understanding.

These debates lead directly into Chapter 12, a discussion of some of the ethical issues surrounding modern genetic research. As the gene is increasingly understood and applied to various modern issues, many ethical dilemmas have arisen for which there are no simple answers. Chapter 12 introduces some of the most important of these issues and, in the process, considers how society can best harness knowledge of genetics while minimizing social harm.

The gene—the molecule of deoxyribonucleic acid (DNA), the coding unit of inheritance, the blueprint for the development of life, the "master molecule"—has attained today an iconic status that few scientific objects share. Much modern culture today ascribes to the gene an almost magical status, capable of explaining almost everything about life. In some ways, modern culture depicts the gene as, according to one commentator, "the secular equivalent of the human soul"—all-powerful, mysterious, and the locus of our individuality. Of course, scientists today do not think of the gene as that mysterious. They have probed its secrets and discovered a molecule that they can isolate, study, and manipulate; today, genetic engineering has brought us cloned sheep, foods containing new genes taken from bacteria, and countless experimental organisms containing genes that they would never have gotten through traditional biological breeding mechanisms. Scientists today know what a gene is. Or do they? Now more than ever, the concept of the gene is at the center of some of our most basic questions about who we are and what we want as individuals and as a society. These issues are complex and important, and one purpose of this book is to provide a historical framework for addressing them. But, beyond this, this book emphasizes how a historical approach to studying science can illuminate the process of science itself and explain the origins and evolution of our modern concept of the gene.

ANCIENT AND MEDIEVAL INHERITANCE

INTRODUCTION

Genetics is in some ways very old and in some ways very new. As we will see later, *genetics* as a scientific discipline began about a hundred years ago, but for most of recorded history people have wondered at the uniqueness of life and at the surprising ability of living things to grow and reproduce in a remarkably consistent way. Oak trees produce acorns that grow into other oak trees; plants (almost) always grow right-side up with their roots in the ground and their leaves in the air; animals grow, change, and produce more of their kind with a remarkable consistency; and family members resemble one another. Life seems ordered, planned, and at least somewhat predictable in its growth, development, and classification into groups, and the mystery of life has been a constant source of wonder and curiosity.

While general comments about life and its mysteries have been found in the earliest writings discovered, the earliest detailed and systematic attempts to explain life have been found in the ancient philosophical texts of ancient Greece. Ancient Greek philosophy is important to the history of the gene because it is important to all Western history; it is the foundation of later Western thought. As described later in this chapter, modified versions of the ancient Greek writings ended up in the hands of Western scholars in the Middle Ages through a complex chain of events, and Greek studies then dominated Western thought for more than five hundred years.

THE LEGACY OF ARISTOTLE

During the Classical (fifth–fourth centuries B.C.) and Hellenistic (fourth–first centuries B.C.) periods, Greece was the dominant culture of the

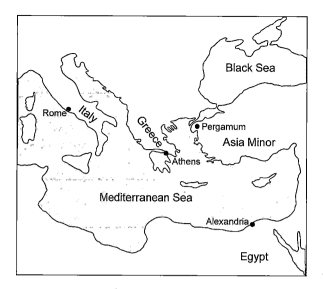

Figure 1.1: Map of ancient Greece. Illustration by Jeff Dixon.

Figure 1.2: Aristotle.

Mediterranean Basin, and Greek influence, from politics, law, literature, religion, art, and philosophy, spread throughout the area and became the foundation of later Western civilizations (Figure 1.1). By far the Greek philosopher who was most important in influencing Western science was Aristotle, a legendary figure in the history of Western thought (Figure 1.2). Much of what we know about ancient Greece we have learned from Aristotle's approximately 30 surviving manuscripts (of a total of approximately 150 that he is thought to have written—Western history might have been profoundly different if only more of these manuscripts had survived). Aristotle was born in northern Greece, in the town of Stagira (now Stavro, a small town in Macedonia), and was educated as a physician. He was sent to Athens in 367 B.C., at the age of 17, to study philosophy. He studied under Plato, an eminent philosopher in his day and still revered in our day. Aristotle studied with Plato for 20 years, becoming a brilliant philosopher in its own right. On leaving Plato's Academy, Aristotle traveled in Asia Minor (modern Turkey) and returned to Macedonia to tutor the son of Alexander the Great; when Alexander conquered Athens, Aristotle returned there and established his own school, the Lyceum, in 355 B.C., where he produced many of his most important writings.

Aristotle disagreed with his teacher Plato in some fundamental ways. Plato believed that the changing world that we see is an illusion, an imperfect copy of another unchanging world—a more fundamental reality—to which we have access only through thoughtful reflection and careful reasoning about the ideal forms that underlie visible reality.

In contrast, Aristotle believed that the world we see is the true reality and that careful observation of the world around us leads to true knowledge.

Aristotle practiced what he preached: his writing touched upon every conceivable subject that humans could experience, from the movement of the tides to the geological formations of the earth, from the purpose of the parts of the smallest living things to the nature and design of the universe. Aristotle described many aspects of biological structure and function, and his writings have also allowed us to reconstruct (albeit imperfectly) some of the major debates about life that dominated ancient Greek thought (and many continued to do so throughout most of the history of the study of life). One of the most important debates involved the nature of living things. What is life? Is there something special about living creatures that makes them different from nonliving objects?

As described in the opening sentences of this chapter, life certainly seemed special to many of those who studied it. Living things seemed to appear out of nowhere, grow, develop, and reproduce copies of themselves with remarkable consistency and orderliness. Many people therefore believed that life was unique and possessed some sort of principle exclusive to living things, which directed its orderly growth, development, and reproduction. Historians refer to these thinkers as vitalists.

But many ancient thinkers disagreed that life was special and opposed the idea of some sort of vital principle unique to life. Instead, these people believed that life was fundamentally no different from nonlife—that, for example, trees and animals were fundamentally the same as rocks and water, except that life had a more complicated structure that resulted in its peculiar characteristics. Those who held such beliefs have been called mechanists. Mechanism and vitalism remained as dominant competing views for about two thousand years.

One of the most influential mechanist camps in Aristotle's time were the atomists, who believed that all the matter in the universe, including all of life, was composed of atoms moving around within a void. These atoms were believed to be indivisible, indestructible, and of infinite size and shape—not the same sort of atoms we think of today, although the modern concept of an atom is in fact descended from the ideas of these ancient atomists. For atomists, life could be explained solely with reference to these atomic units of matter in motion, combining and separating in complex ways to produce, given enough time, all the characteristics of life. There was no need to imagine some sort of mysterious quality unique to life in order to explain it—life was simply a particular arrangement of atoms that, somehow, resulted in life's peculiar qualities. The most important atomists of ancient Greece were Democritus, active in the late fifth century B.C., and Epicurus, a contemporary of Aristotle.

Another ancient mechanist was Empedocles, who had a theory similar to that of the atomists but slightly more complex. Empedocles wrote that all

The Importance of Vitalism

As we will discuss in a later chapter, the ancient philosophy of mechanism experienced a revival in the seventeenth century with the rise of the mechanical philosophy, promoted by the French philosopher René Descartes and others. Mechanical philosophers argued that life is best studied as if it were a machine and that philosophers should avoiding seeking the purpose of life's various functions and activities. Like the early Greek atomists, they dismissed such topics as too mysterious or occult and as inappropriate as an aid for understanding the natural world. The mechanical philosophy resulted in a great deal of success, and much of modern science owes a debt to the view. Many historians of science once viewed the rise of the mechanical philosophy as the foundations of modern science.

But the mechanical philosophy did not lead to the early-twentieth-century understanding of the function of the chromosomes, which we now know are responsible for heredity. Instead, early research into chromosomes was motivated by a revival in nineteenth-century Germany of vitalism, which argued, much like Aristotle, that there is something unique about life that explains its special properties. Therefore, far from dismissing it as occult, we must recognize that vitalism is connected to two of the greatest discoveries of the twentieth century—the chromosomal theory of inheritance and the discovery of the structure of deoxyribonucleic acid (DNA). Even today DNA is often described as the "secret of life," which sounds much more like the language of the vitalists than the mechanists. Chromosomes and DNA are discussed in much greater detail in Chapter 7.

matter could be explained by four elements, earth, air, fire, water; and two forces, love and hate, or attraction and repulsion. These four elements, under the influence of love and hate, combined and separated to form all of the universe and all of matter. According to Empedocles, life in particular resulted from a peculiar, random combination of parts of matter during a certain time period in the history of the universe—the Age of Love—when the force of attraction controlled the forms of matter. As a result, body parts and organs simply got stuck together randomly; some of these combination produced monstrous atrocities that could not function, but occasionally functional forms were created. Those combinations that were functional— that were well adapted to survival— remained as permanent animals and plants, produced offspring, and increased in numbers.

Therefore, both Empedocles and the atomists were mechanists, because they believed that life could be explained using basic concepts from physics—exclusively matter in motion for the atomists, while Empedocles added his two forces of attraction (love) and repulsion (hate).

Unlike Empedocles and the atomists, Aristotle was not a mechanist. He did not believe that such random events as chance combining of parts and matter in motion could explain why organisms are formed, grow, and function in such a consistent manner. He argued that those who study nature and life could not simply suggest some mechanical process that resulted in its shape and functioning. Instead, Aristotle argued, one should always include how and why this structure has the function it does. What is the purpose of this or that part of an animal? Why is a part arranged in a certain way, and what does it do? Aristotle's emphasis on purpose is an example of what philosophers call *teleology* (from the Greek word *telos*), which means to explain something by referring to its final purpose. For example, a teleological explanation for why a giraffe has a long neck would suggest that the neck allows the giraffe to reach the top of tall trees.

For Aristotle, questions about purpose and function were an essential part of the explanation of life. For the atomists, such questions were irrelevant, because they went beyond a straightforward mechanical description. For the atomists, there is no purpose: life simply resulted from random mundane events. Discussion of purpose sounded too mystical and mysterious and therefore had no place in discussions of the natural world. Even discussion of function was seen as moving away

Empedocles and Darwin

Empedocles believed that, during the Age of Love, various body parts combined and that those that were best able to function survived. Does this remind you of anybody else's theory? Later in this book we will discuss Charles Darwin, the nineteenth-century naturalist who formulated his famous theory of natural selection. On the surface, Empedocles' theory sounds similar to Darwin's, with its discussion of random forms that are "naturally selected" if they function well (note that Empedocles did not use the phrase "natural selection"). But Empedocles' theory is in fact quite different because, unlike Darwin, Empedocles was not trying to explain evolution. Empedocles argued that this process of randomly combining parts and the survival of those that functioned well happened at only one time in the distant past. After the Age of Love, the creation of living forms was over, and the forms that survived the process are the ones that remained, unchanged.

from a simple mechanical explanation. For example, a human hand might move in a certain way, but it has no intrinsic function: this sounds too much like it was designed for the purpose of that function. Instead, the fact that a hand can grasp things is, for the atomists, not really part of a proper understanding of the hand.

Aristotle's notion of teleology was that the purpose of a thing was to be found in the form of the thing itself; purpose was thus intrinsic, or internal, to the thing. This has traditionally been referred to as internal teleology. In this respect, Aristotle differed fundamentally from Plato, his teacher. Because he believed that the natural world was merely an imperfect reflection of true reality, Plato argued that true purpose was found not internal but external to a thing. For Plato in particular, the purpose of a thing could be found in the mind of the creator of the true, ideal form of the thing. Plato called this creator the *demiurge,* a God-like being with access to the "true reality" that consisted of ideal forms, which the *demiurge* used to create the physical world. As we will see later in this chapter, external teleology became an important component of religious explanations of the natural world in the Middle Ages: medieval thinkers adopted much of Aristotle's work but reinterpreted it as an example of external, rather than internal, teleology.

Aristotle's teleology had multiple components. He argued that there are four causes to be found acting in the natural world and that an explanation or description of, for example, a particular animal or a particular body part must include all four of these. "Cause" in this sense is best understood as "components of an explanation." Thus, Aristotle believed that a proper explanation, description, or understanding of life has the following four components.

The first is material cause, which refers to the material from which a living thing is made; the second component is the efficient cause, which refers to the process by which a living thing comes into being—that is, *how* life is made; the third is the formal cause, which refers to the form—the shape, or structure—of a living thing; and the last is the final cause, which refers to the reason, function, or purpose of a living thing (such as a body part). Aristotle used analogies to explain his four causes. For example, imagine that we wanted to explain a statue using Aristotle's four causes. In this case, the material cause of the statue is the stone, or marble, or other material from which the statue is made; the efficient cause is the process of its construction—that is, presumably the original stone was carved using some sort of tool; the formal cause is the shape of the statue itself; and the final cause might be to commemorate somebody, to create a work of art, or perhaps to make money.

However, and as Aristotle himself argued, there are problems with this sort of analogy when applied to living things. The efficient, formal, and final causes of the statue are *external* to it: there is a creator (the sculptor), separate from the statue, who presumably has a form in mind and a purpose for the statue. This is an example of *external teleology,* the view that the purpose of a thing is external to the thing itself. As we have seen, Plato, Aristotle's teacher, believed in external teleology; he believed that all sensible objects were created by a deity he called the *demiurge,* which existed in a world of forms and which used these forms to create the objects of our world. In contrast, Aristotle believed in *internal* teleology, arguing that all four causes of an organism are *internal* to it. The plan or form of the organism, the way that it is made, and its purpose are all a part of the organism, with no existence independent of the thing itself. For Aristotle, there is no "sculptor of life": not just the material but the form, process, and purpose are all inherent to the organism. As will be seen later, Aristotle's description of teleology was often misinterpreted; for example, in the Middle Ages, Aristotle's ideas became the foundation of much of Christian thought, but in the process his teleological arguments were seen as evidence for the existence of a Christian God.

Aristotle used the four causes to explain many things, including the heredity and development of an offspring. One of the major questions for those who studied life was the question of organic *generation,* how an organism takes shape and develops as a product of their parents. *Generation* was a concept that included both heredity and development. It included how parents pass on traits to an offspring and how the offspring then develops into a fully formed organism (as discussed later, heredity and development were seen as part of the same process, and referred to as generation, until the early twentieth century).

How did atomists explain why offspring looked like their parents? Atomists argued that semen from the male contains small bits of matter from every part of the male's body. When semen is transferred to the female, the atomists argued, these tiny bits of matter recombined to form the offspring. The details of this recombining are unclear, partly because much of what we know about the atomists we know from Aristotle's critical description of them.

Aristotle argued that atomism could not explain why the offspring of organisms always get assembled properly. How is this recombining of tiny bits of matter guided and directed? Something must *cause* all the parts to assume the correct form. It cannot be just random, because the correct form almost always gets made, and this form happens to have a structure that is extremely well suited to its actual function. Aristotle thus saw in the hereditary process an elegance and a consistency that went far beyond the explanations offered by the atomists. The chief issue for Aristotle, then, involved explaining how *form* (that is, formal cause) could be transferred from parents to offspring. Aristotle's answer was that the division of the formal and material causes was reflected in the division of the sexes into male and female. Aristotle argued that individuals are assembled from menstrual blood and that females therefore supplied the matter (material cause); males, in contrast, supplied the form. The male semen, according to Aristotle, carries the formal cause within it and impresses this form onto the menstrual blood, the material from which an organism is made, in order to produce an offspring. As is perhaps obvious, Aristotle was heavily influenced by the bias against women that was common in Greek culture.

While generation was of prime concern for Aristotle, he also covered many more issues about the nature of life. His writings on anatomy, physiology, animal classification, and other topics dominated the study of life, well into the sixteenth century. For Aristotle, the study of living things was a perfect example of how observing and explaining the richness of the natural world illustrates his fundamental views of a universe governed by causes.

GALEN OF PERGAMUM AND ANIMAL GENERATION

Approximately two hundred years after the decline of the Greek Empire in the third century B.C., the Roman Empire arose and dominated much of the Mediterranean for about the next five hundred years. Greek thought and language, however, remained very influential in Mediterranean centers of learning, especially in Rome itself but also Pergamum in Asia Minor and Alexandria in Egypt, both of which were in fact Greek-speaking regions throughout much of the history of ancient Rome. Aristotle remained very influential, and his treatises on living things were taken up and modified in various ways.

One very influential individual, who was himself heavily influenced by Aristotle, was Galen of Pergamum, a philosopher and medical practitioner whose views on human and animal anatomy and physiology dominated medical practice for more than 1,500 years (Figure 1.3). Galen studied medicine in several centers in Greece and then returned to his home in Pergamum at age 28 to become physician to the gladiators. A few years later, in 162 A.D., Galen moved to Rome to serve as personal physician to the Roman emperor Marcus Aurelius and Aurelius's son, Commodus. Galen was thus very important and influential both in his own time and after his death. Galen's ideas dominated medicine, anatomy and physiology until the seventeenth century, although they were under increasing attack by the sixteenth century. Galen was greatly influenced by Aristotle's insistence on observation and expanded much of Aristotle's own observations of animal structure and function.

Galen was also greatly influenced by Aristotle's teleology. Like Aristotle, Galen was a vitalist; he argued that life is special and unique and shows evidence of purpose. In much of his writing, he criticizes the views of atomists, indicating that the atomistic view remained popular even after Aristotle criticized it five hundred years previously. Atomism eventually declined in popularity under Christianity, likely because of its mechanistic and purposeless nature. Despite its internal teleology, Aristotelianism still offered more room for a soul, human free will, and divine creation than did the atomistic universe.

Galen performed a great variety of experiments to demonstrate that animal parts were full of purpose; Galen called these *natural faculties,* inherent properties that explain

Figure 1.3: Galen of Pergamum. Courtesy of the National Library of Medicine.

why they behave as they do. Like Aristotle, Galen believed that these properties were an important part of the explanation of life and that natural faculties illustrated the final cause of the body parts to which they belonged. In 170 A.D. Galen published *On the Natural Faculties* to describe these functions and illustrate this argument.

However, and unlike Aristotle, Galen argued that there was indeed a Divine Creator, who had structured living organisms to be so well suited to their function. His was an external teleology, as opposed to Aristotle's internal teleology. For Galen, Aristotle's analogy of the sculptor who created a statue exactly describes the creation of living organisms by what we might call a "divine sculptor." Galen believed that this external creator gave organs their natural faculties.

Galen also was very influenced by Aristotle's views on heredity, but again in this case he differed from Aristotle. More specifically, he disagreed with what he viewed as a flawed characterization of Aristotle's views on heredity. In his (ca. 180) treatise *De Semine (On Semen, or On Seed)*, Galen argued that the standard interpretation of Aristotle's view by other commentators of Galen's time, that the male provides the formal cause and the female provides the material cause, is in fact a misreading of Aristotle. Galen argued that Aristotle actually believed that *both* parents provide a formal cause. For Galen, it seemed self-evident that both parents provided a formal cause to their offspring, because of the empirically observable fact that children as often as not resemble the mother as much as the father. The "one seed" theory in which the father's seed alone fashions the child cannot explain this observation, except to say that offspring sometimes deviate from the paternal form by accident. For Galen, this was absurd, since one can observe this sort of "accident" more frequently than one can see pure paternal resemblance. More often than not, Galen argued, one observes a mixture of paternal and maternal features, and for this reason he believed that both parents contained a seed that somehow combined to provide formal cause to the offspring.

Galen approaches the issue as a true Aristotelian, by conducting anatomical observations. He performed extensive dissections of female apes. He observed a fluid in the horns of the uterus, which to him was analogous to the male semen; he thus believed that he had discovered female seed, the source of the female's formal cause. While he was mistaken, Galen's insistence on female formal cause and his attempts to explain it are remarkably insightful from the perspective of our contemporary knowledge of heredity. Additionally, his commitment to testing his theory with careful observation is an example of the sort of clear and systematic work that helped make Galen the prime medical authority for much of the history of science.

The most famous and important example of Galen's contributions to anatomy and physiology is his detailed theory of blood flow, in which he described two separate blood systems having two separate purposes, one centered around the liver that nourished the body and another around the heart that provided "vital spirit," a life-giving force that gave the body vitality. This theory remained one of the foundations of medicine until challenged in the sixteenth century and eventually overturned by William Harvey's discovery of blood circulation in the seventeenth.

HEREDITY IN THE ISLAMIC EMPIRE

As described earlier, Roman scholars under the Roman Empire were the first to use and modify Greek thought; Arabic scholars under the Islamic Empire were the second. From the second to the sixth centuries A.D., the Roman Empire declined, starting with its division into an Eastern, Greek-speaking Byzantine Empire and a Western, Latin-speaking Roman Empire. Whereas the upper classes of Roman society at the height of the Roman Empire knew Greek, after the empire's decline few people did, and access to Greek thought became almost nonexistent. Christianity became very powerful in both the Greek and the Latin parts of the former empire, and education and scholarship focused almost exclusively on Latin Christian texts. Aristotle in particular was dismissed by many early influential medieval scholars, such as Saint Augustine. Augustine was a very influential theologian and thinker who viewed Aristotelianism as pagan philosophy, rendered obsolete by Christian thinking. Augustine did, however, make use of Platonic philosophy, which was commonly viewed as consistent with Christianity because of its description of an ideal world beyond the illusory physical world. But the works of Aristotle became lost to the West during the early Middle Ages.

But by the seventh century, following the death of Mohammed, an Islamic Empire rapidly rose to prominence in the Mediterranean, and it is here that Greek art, culture, and philosophy were used, modified, and thrived for more than five hundred years. The Greek language and Greek culture was well known within early Islamic culture, partly because Greek Christian missionaries from the Byzantine Empire often had successfully established Christian settlements in parts of the East. Additionally, the Islamic Empire itself included many former Greek-speaking areas. This mixture of cultures resulted in an important period of translation of Greek texts into Arabic. The height of this translation period was in the eighth and ninth centuries. Translated texts were gradually distributed throughout the Islamic Empire, forming an important foundation for Islamic philosophical and artistic thought. Between approximately 900 and 1200 A.D., many important commentaries and original philosophical works were written that became almost as important to Islamic and medieval scholars as the original Greek texts.

Averroes (or ibn-Rushd) was a famous and influential court physician and philosopher and the most important of the Islamic Aristotelian scholars

(Figure 1.4). From more than fifty surviving texts we know that Averroes wrote
on philosophy, theology, medicine, astronomy, and law, and his writings were
very influential in Arabic and, later,
Western medieval philosophy. By the
time Averroes was producing his writings, Aristotle had become influential in
the Islamic world, but Averroes differed
significantly from some Islamic scholars who criticized Aristotelian writings
as inconsistent and incompatible with
Islam. In contrast, Averroes' goal was
to reconcile the two. He believed that,
when properly understood, philosophy
and religion were compatible. Averroes'
writings on Aristotle spanned more than
twenty-five years and included commentaries and treatises on most of the Aristotelian texts. Averroes' writings were so
influential that, hundreds of years later,
he was often referred to as "the Com-

Figure 1.4: Averroes. Courtesy of the National Library of Medicine.

mentator," while Aristotle himself was often called "the Philosopher."

Averroes' writings on heredity and generation were similar to those of Aristotle's,
but at the same time they caused difficulty for him in his attempts to reconcile
Aristotle with Islam. Averroes, like Aristotle, saw heredity as the transmission of
formal and material cause from the male and female, respectively. However, he
criticized Aristotle's conception of formal cause for leaving no room for the notion of an immortal soul that survived the physical body. Averroes' criticism was
partly political: he believed that people *needed* the concept of the immortal and
immaterial soul in order to live good and ethical lives. He also appears to offer a
metaphorical conception of the soul as residing in the general form of a species,
rather than in any particular individual. This might appear to suggest that Averroes truly did not believe in the existence of an immortal soul, but his argument
is more subtle and complicated than this. He in fact believed that there was "religious truth" in the idea of the soul; Averroes argued that "religious truth" differed
from "philosophical truth" but that neither was superior to the other.

Averroes' writings on heredity and generation are an excellent example of
some of the difficulties he had reconciling Aristotelianism and Islamicism. At
the same time, they reveal how his religion motivated his studies, and more
generally they are an example of the complexities of interaction between scientific and religious beliefs in the history of science. Many popular descriptions of science and religion depict the two as being perpetually at war, with

science being portrayed as trying to banish the superstitious understandings of the world offered by religion. More often than not, however, individuals in the history of science were motivated by both religion *and* science to produce some of their greatest works. As we will see, this is true not only in the case of Averroes but for many later scientists up until even the nineteenth century.

The importance of Islamic influence on both Greek and Roman natural philosophy cannot be overstated. Much of the Greek corpus that we know today is in fact a combination of Greek, Roman, and Arabic thought (although Galen was in fact modified very little). Yet, for much of the history of science, this importance was unacknowledged. There have been several reasons suggested for this. One important reason is that many early Western historians of science looked at Islam as basically an interruption of Greek and Roman learning, until this lineage was restored with the rise of the West. The Arabic world was seen not as an important intellectual center in its own right but rather as a holding station, preserving ancient texts until the West was ready to receive them. Civilization was seen as declining after the fall of Rome and reviving in the later medieval period. This was true enough for the Latin world and the rest of the West, but no such break could be seen in the Mediterranean basin, where cities and intellectual life continued to thrive, but under Islamic rather than Greek or Roman influence. This Western perspective on Islamic knowledge ignored the importance of ancient thought in Islamic life itself, in its own right. This skewed perspective has been modified somewhat in the past few decades, but remnants of this attitude often still affect how the history of science is portrayed and taught.

HEREDITY IN THE MEDIEVAL WEST

For about five hundred years, the Islamic world was the center of intellectual activity around the Mediterranean basin. Meanwhile, areas north and west of the Mediterranean began to be populated by nomadic Germanic tribes, and these groups eventually began to adapt to the harsher climate and land of northwestern Europe. From the tenth to twelfth century, the widespread massive use of new technologies such as the plough and the horse bridle and of new agricultural methods such as crop rotation allowed for the creation of agricultural communities, which gradually led to larger urban regions. As these regions of Europe began to become more viable areas of living, they, like those before them, gradually developed a culture of intellectual thought. At the same time, Christianity grew in power and influence, especially through the growth of the Holy Roman Empire from the tenth to the seventeenth century.

By the twelfth century, Western Europe began to obtain access to ancient Greek, Roman, and Islamic writings through translations of these texts from Arabic, Syriac, and sometimes Greek into Latin. As the West became exposed to these important intellectual traditions, Western philosophers adopted them and adapted them to their own contexts. The result was that once again the center of learning shifted, this time to northwestern Europe.

Another important medieval European development was the growth of universities—autonomous associations similar in organization to medieval craft guilds (in fact, the Latin word *universitas* traditionally referred to any collection of professionals). Medieval craft guilds had a hierarchy of apprentices, journeymen, and master workmen; the universities adopted a similar structure of matriculation, bachelor, and master (the teacher). But universities were unique, and nothing exactly like them had ever existed previously. Members were (with some exceptions) especially privileged and were given leisure time to study knowledge not necessarily immediately relevant to practical aims. Universities were pivotal to the later dramatic rise in prominence of science in the West (although rising European economic and political power and, later, imperialist expansion and colonialism are also important foundations for both universities and the rise of Western science). By the thirteenth century, universities were established in centers in Italy (Padua and Bologna), France (Paris), and England (Oxford and Cambridge). The Italian universities remained the most powerful and influential institutions for hundreds of years. Later, universities arose in Germany and then in other areas of Europe, and in the process Classical and Arabic philosophy were disseminated throughout Western Europe.

The comprehensive study and teaching of ancient philosophy were rare in Europe until the rise of universities, but early Christians had studied and taught some Greek writings in monasteries and other institutions as an aid to understanding Christian scriptures. Christianity remained central to the new universities, as well: several religious orders, most notably the Dominicans and Franciscans, encouraged the development of universities and scholarship as part of their missionary efforts. Theologians were typically also philosophers and became experts in Aristotelian philosophy in particular, while adapting it to the Christian religion. This Christian interpretation of ancient Greek and Arabic philosophies became referred to as Scholasticism, a mixture of theology and philosophy that was the chief domain of intellectual thought in the Middle Ages.

One of the most important of the Scholastics was Saint Thomas Aquinas, who regularly used "the Philosopher" and "the Commentator," as he frequently called Aristotle and Averroes respectively, in his attempts to understand and interpret Aristotle within a Christian framework. Aquinas was a Dominican theologian and philosopher at the University of Paris, who was strongly influenced by Aristotle and who was arguably the most important theologian to contribute to the rise of Scholasticism.

Aquinas was particularly opposed to Plato's belief in forms external to physical objects and followed Aristotle in his belief that formal cause is internal to objects themselves. But generally Aquinas did not express much interest in Aristotle's observations of nature, and instead he was more interested in his metaphysics and his proof of the Prime Mover as the ultimate cause of the universe. In his metaphysical writings, Aristotle had invoked the Prime Mover as the ultimate cause of all activity in the universe. Everything in the universe happens as a result of some sort of cause, but how did all of these events and activities first get set in motion? According to Aristotle, there must be a first, or Prime, Mover that started all of the activity of the universe. Aristotle thus saw the study of nature as proof of this Prime Mover, an eternal, immaterial being that is the ultimate, final cause of all of reality.

Aquinas argued that Aristotle's metaphysics also illustrated the notion of the immaterial human soul, whatever Aristotle had said about internal teleology. Thus (and to put it rather simply), by focusing primarily on metaphysics, which was much more amenable to Christian theology by describing God as the Prime Mover, rather than natural philosophy, with its notion of internal teleology, Aquinas revealed compatibilities between Aristotle and Christianity, and as a result he greatly contributed to the influence of Aristotle within Christian teachings.

CONCLUSION

While modern genetics is in fact a very new discipline, its roots are thousands of years old. Our oldest examples of recorded history illustrate the age-old fascination with living things that display a remarkable regularity in their growth and reproduction. Ancient Greek philosophy is our earliest example of a systematic attempt to understand the nature of this regularity. Aristotle believed that life has some sort of unique principle that accounts for its remarkable capacity to faithfully pass on its form and functioning to later generations. His notion of formal cause was the pillar upon which many subsequent generations of philosophers explained the phenomenon of *generation*, the hereditary and developmental properties of life. Beginning with Galen, during the height of the Roman Empire, continuing with Averroes, under the Islamic Empire, and proceeding to early Christian Scholasticism, developed in medieval universities by scholars such as Thomas Aquinas, Aristotle's philosophy remained at the foundation of human understanding of nature until it was challenged during the Enlightenment, as described in the next chapter.

2

RENAISSANCE AND ENLIGHTENMENT VIEWS OF INHERITANCE

INTRODUCTION

In Chapter 1 we witnessed the influence of Aristotelian views of *generation*, heredity and development, up until the rise of medieval Western university scholarship under the Scholastics in the twelfth and thirteenth centuries. By the thirteenth century, Scholasticism was the dominant tradition of learning in universities. But, for the next several centuries, increasingly detailed studies of nature, new technologies and institutions, and a great increase in the production of written materials led to increasingly sophisticated and aggressive challenges to ancient thought, and especially to the Scholastic Aristotelianism of the universities.

But the reaction against Aristotelianism produced a problem for studies of living things, and in particular it became increasingly difficult to understand processes such as heredity and development. The question of how offspring were formed and developed was a topic of increasing debate, set within the context of broader debates about the value of Aristotelianism and the proper way of understanding nature. As we will see, while the seventeenth century saw the rise of a strict mechanical account of the world, reminiscent of the views of the ancient atomists, the problem of generation led once again to an argument similar to that which Aristotle used against the atomists: life had some sort of uniqueness to it for which mechanism could not account.

GROWING CHALLENGES TO SCHOLASTICISM

As medieval Europe expanded, so too did attempts to understand the natural world, and in this context the traditional Scholasticism of the universities came under increasing scrutiny. A variety of developments contributed to this. The invention and widespread use of the printing press in the mid-fifteenth century resulted in a subsequent increase in access to texts among the upper- and upper-middle-class men of Europe; humanism, a movement to encourage the "purifying" of ancient texts, later led to the extension and modification of ancient knowledge and, eventually, greater criticism; the medieval Christian church experienced increasing criticism and challenges, such as the Protestant Reformation of the sixteenth and mid-seventeenth centuries; increasingly sophisticated technologies and craft traditions produced skepticism regarding traditional Aristotelian knowledge and its perceived irrelevance for practical needs; the rise of *colonialism*—European expansion into and exploration of other parts of the world—brought many new discoveries that expanded the known world for the Europeans; and developments in traditional areas like medicine and astronomy, while not exactly rejecting ancient thinking, created room for criticism and a culture of independent observation within the field of natural philosophy.

NATURAL PHILOSOPHY

Natural philosophy was the term used to describe the study of nature for most of the history of such studies. The term originated in the thirteenth century, and was used until the eighteenth century to describe scholarly studies of the natural world. Our modern conception of a scientist is approximately equal to that of the natural philosopher, but there are some crucial differences. Natural philosophy was first and foremost about God. With its role of exploring the relationship of God to the universe, natural philosophy has been described by some historians as a "handmaiden" to religion. The primary goal of a natural philosopher, then, was to explain the nature of the entire universe and to understand the universe in relation to God. All natural philosophers shared this goal. Traditionally, many historians described many of the historical figures described in this chapter—William Harvey, René Descartes, and Isaac Newton, for example—as trying to fight against religion and to instead use "science" to understand the world. A famous example often used to demonstrate this conflict is the trials of Galileo, in which the Church condemned Galileo for creating a model of the solar system in which the sun, rather than the earth, occupied the center. But in fact Galileo believed in God and did not see his system as atheistic. On the contrary, he saw his explanation of the universe as simultaneously an explanation about its relationship to God. Similarly, as is described later, many natural philosophers' descriptions of the universe were very much about understanding the role of God in that universe, because understanding that relationship was the primary goal of their scholarly efforts.

While initially there were disagreements about the extent to which religious scholars should study natural philosophy rather than studying the Bible itself, by the fifteenth and sixteenth centuries, the field of natural philosophy was an established practice, and scholars began to study the natural world with increasing sophistication (as discussed in the preceding chapter). But the central goal of natural philosophy remained an understanding of God and the relationship of the universe to God's existence.

CHALLENGES TO ARISTOTELIANISM

By the seventeenth century, Europe had greatly expanded, had many urban centers, and had engaged in trade with many parts of the world. Europeans had seen their world expand and change dramatically within a period of just over a hundred years. Study of the natural world also changed considerably. Scientific societies were formed, such as the Royal Society of London, the Académie des Sciences in Paris, and the Accademia del Cimento in Tuscany; these provided alternative institutions to universities, which were dominated by Scholasticism. In these institutions, various seventeenth-century natural philosophers offered increasingly specific critiques of Scholasticism and Aristotelianism. In his 1620 book *Instauratio Magna* (The Great Instauration), Francis Bacon criticized the traditional Aristotelian practice of simply observing nature in its undisturbed state (Figure 2.1). Bacon, in contrast, insisted on the use of tools and experimentation in order to probe nature and uncover its secrets. For an Aristotelian, to disturb nature was to disrupt its natural essence and therefore to not understand it at all, but for Bacon, this was the key to obtaining knowledge. Bacon also argued for a

Figure 2.1: Frontispiece of Francis Bacon's *Instauratio Magna*. Image copyright History of Science Collections, University of Oklahoma Libraries.

unified approach to describing nature and the importance of using knowledge of the world in order to exhibit control over it.

Other seventeenth-century developments challenged the classical worldview. William Harvey, in his 1628 book *De Motu Cordis* (On the Motion of Blood), revolutionized knowledge of human anatomy when he presented the process of blood circulation, overthrowing the Galenic theory of blood flow that had persisted for more than a thousand years. Finally, the flourishing of microscopic studies during the seventeenth century revealed an enormous new world of living things and encouraged a growing view that the world of the ancients was only a subset of the entirety of nature. By the late seventeenth century, these developments and more led to a revolution against Aristotelianism, affecting all areas of natural philosophy.

THE MECHANICAL PHILOSOPHY

One very influential seventeenth-century development was the rise of the mechanical philosophy, advocated by a variety of natural philosophers, the best-known being René Descartes (Figure 2.2). In his 1637 *Discours de la Méthode* (Discourse on Method), Descartes argued that nature was a system of interconnected machines: living things, including humans, could be described with respect to the same natural processes that one could see in the workings of inanimate, mechanical objects. The mechanical philosophy was in some ways a revived and revised atomism like that of the ancient Greek opponents of Aristotle and Galen. Like the atomists, mechanical philosophers believed that the activity of nature was the result of random motion and interaction of particles, and they excluded any explanation for the workings of the universe that invoked purpose.

Historians once thought that the mechanical philosophy was in direct conflict with religious explanations offered by traditional Scholasticism, but this is not quite accurate. Like all natural philosophers, the primary goal of the mechanists was to understand the relation of the universe to God. The mechanists described this relationship by arguing that God created the

Figure 2.2: René Descartes. Image copyright History of Science Collections, University of Oklahoma Libraries.

universe but did not intervene in it; instead, God allowed the universe to function according to the mechanical laws that He himself put into place. Additionally, mechanists believed that there was much about nature that they could not explain, but for them this simply meant that such explanations were the role of theology, rather than science. Descartes, for example, described the human body as if it were a machine, but he excluded the human rational mind from his mechanical philosophy. While some mechanical philosophers attempted to explain the working of the mind mechanically, others viewed the mind as outside the domain of natural philosophy and instead belonging to theology.

Advocates of the mechanical philosophy also argued that Aristotelian attempts at understanding form and function in life should also be left to theologians, not because they were dismissive of form and function but because they thought that the elegance and beauty of form and function could be explained only by reference to a divine Creator. Mechanists themselves thus believed that God created life so as to be well suited to its environment and left the details of this to theologians. This belief that the intricacy of form and function was a result of divine action was popularized by writers of natural theology, a late-seventeenth-century movement that described nature's complexity as evidence of God's perfection. This argument was often referred to as the "argument from design." According to the argument from design, life was so intricate and complex that, clearly, it had to have been designed by a Creator (modern "intelligent design" advocates make similar claims). One of the best-known natural theologians was John Ray, who in 1691 wrote *The Wisdom of God Manifested in the Works of the Creation* (Figure 2.3).

Such writings complemented the mechanical philosophy; in fact, mechanists themselves made the same argument from design to exclude Aristotelian explanations of form, purpose, and function from the domain of science. The microscopists, for example, who were mechanists, argued that their discoveries also

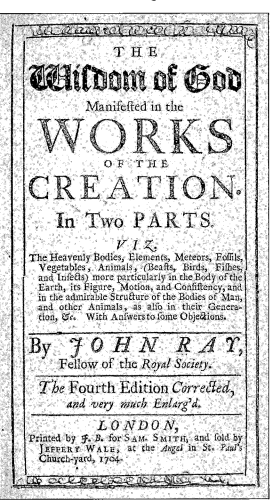

Figure 2.3: Front page of John Ray's *Wisdom of God Manifested in the Works of the Creation.* Image copyright History of Science Collections, University of Oklahoma Libraries.

supported the view of the elegant fit of form to function, observed even in the smallest of creatures, and this gave further support for the magnificence and perfection of God.

MECHANICAL ENPLANATIONS OF HEREDITY

The mechanical philosophy held that all phenomena observed in living things could be explained with reference to simple matter in motion and simple mechanical effects on matter. Life, too, could be explained as different sorts of matter physically contacting and thus affecting each other. For the mechanists, there was no need to resort to what they called "occult" qualities such as, for example, Aristotelian formal and final causes. This included processes of generation: mechanists had a variety of mechanical explanations to explain how offspring originated and developed. As Descartes said in his *Discourse on Method:* "it is not less natural for a clock, made of the requisite numbers of wheels, to indicate the hours, than for a tree which has sprung from this or that seed, to produce a particular fruit."

As an alternative to Aristotelianism, Descartes had offered in his *Discourse on Method* what many thought was a weak explanation of generation. Descartes, of course, could not accept the addition of a formal cause and sought a mechanical explanation. He suggested that the paternal and maternal materials, when mixed together, caused microscopic particles within these materials to intermingle, to press against each other, and somehow to form the parts of the offspring according to natural, mechanical principles. This is of course a vague description. Descartes argued, however, that if and when natural philosophers understood the precise structures of the particles in the male and female material, they would understand exactly why they interacted to form parts of the offspring. He believed that the process was entirely predictable and mathematically precise, if only one could know the shape of these microscopic particles.

In response to Descartes, William Harvey, who, as described earlier, revolutionized anatomy and physiology with his discovery of the circulation of the blood and his publication, in 1628, of *De Motu Cordis,* published another book in 1651 titled *De Generatione* (On Generation), in which he defended the Aristotelian view that male seed provides an immaterial, formal cause to egg's development. Harvey, despite having overthrown ancient Galenic beliefs about blood flow in the body, was himself an Aristotelian and opposed to mechanism. Harvey's views were influential, and his criticism of Descartes was important: Descartes had himself used Harvey's description of the human heart as acting like a pump to demonstrate that the human body functioned like a machine. *De Generatione* clarified Harvey's opposition to such a conclusion and to mechanism in general.

Few natural philosophers, therefore, were convinced by Descartes' speculations; even other mechanists felt compelled to come up with other ways to explain generation. Animal generation was in fact the most important difficulty for the mechanical philosophers; despite the official mechanist view that explanations for form and function lay outside the boundaries of natural philosophy, by the mid-seventeenth century this lack of explanation was typically seen as a gap in the mechanical philosophy.

By far the most common "explanation" was in fact a theory that allowed mechanical philosophers to *avoid* an explanation. The theory was known as preformation, which stated that all organisms were preformed in their parents. The most common version, called emboitement, was developed by Jan Swammerdam in his 1672 publication, *Miraculum Naturae* (The Miracle of Nature). Swammerdam argued that all organisms that have ever existed and will ever exist are preformed one inside the other within the eggs of females. All life was thus preformed at one point in the distant past, and generation could be explained with reference to this original creation. The moment of creation was an act of God, itself unexplainable by science. The production of form, then, required no explanation according to this theory, because it was the work of God.

Preformation was thus very compatible with mechanism. According to preformation theory, development involved the mechanical unfolding and growth of preexisting forms designed by God, which is in keeping with the belief that God does not regularly intervene in the universe. Additionally, it reiterated the view that the process of acquiring form was beyond the boundaries of science and was instead a theological question. Various individuals in the seventeenth century advocated preformationism, but Nicolas Malebranche, a mechanist who supported Descartes' central ideas concerning the mechanical philosophy, wrote a book in 1674 titled *Search for Truth*, which gave the theory a strong following. Preformation theory provided an answer of sorts to the problem of generation, allowing the mechanists to include living things in their general worldview. The mechanical philosophy was a very conscious challenge to Aristotelianism, which itself was extremely comprehensive; any challenge to Aristotle must therefore be equally comprehensive. As a result, the mechanists were strongly motivated to provide a mechanistic view of life, despite the awkwardness of preformation theory.

But, in fact, several observations were seen as providing evidence for the theory of preformation. Swammerdam himself suggested emboitement in response to his observation of rudimentary nymph structure within a dissected caterpillar and of rudimentary butterfly structure within the nymph. Additionally, the successes of microscopic studies in the seventeenth century, as mentioned earlier, allowed microscopists to see extremely small but complex forms of life. The mechanistic microscopists thus had no difficulty imagining infinitely small life forms, which would be required for preformation to be possible. For example, Antoni von Leeuwenhoek (Figure 2.4) built many excellent microscopes and discovered many microscopic organisms, opening

Figure 2.4: Antoni von Leeuwenhoek. Copyright © Rijksmuseum Amsterdam.

up a whole new world of living things. In 1673, Leeuwenhoek began writing letters to the Royal Society describing his observations, and, in a 1685 letter, Leeuwenhoek described his discovery of what he called "animalcules," tiny microscopic organisms that he claimed to observe in male seminal fluid. Today we describe Leeuwenhoek as having discovered sperm cells, but he himself believed that these were fully preformed animals that would use the female egg as nourishment to grow.

Leeuwenhoek's animalcule theory never really caught on; the most popular preformation theory was that the female egg, or fluid in the womb, contained the preformed offspring, rather than the male semen. Leeuwenhoek was himself not a respected natural philosopher and was not well educated. His observations could not be reproduced. Additionally, Leeuwenhoek's study of male semen was seen as rather improper and distasteful. Leeuwenhoek did not help matters much when, at one point, he insisted that he obtained his male sperm not by "sinful contrivance" but by rushing from his and his wife's bed to the microscope! The entire topic was considered quite unseemly.

Several other pieces of evidence in the eighteenth century were interpreted as supporting preformation. For example, in 1740, Charles Bonnett observed a female louse raised in isolation that gave birth without male involvement (a process called parthenogenesis). Bonnett explained the observation using preformation theory, arguing that, when a male is involved in reproduction, his role is merely to trigger the growth process of the preformed offspring that existed in the maternal womb. In the case of parthenogenesis in the louse, the growth process was triggered in some other way.

HEREDITY'S CHALLENGE TO THE MECHANICAL PHILOSOPHY

Despite its dominance for most of the seventeenth century, in the eighteenth century the mechanical philosophy came under increasing challenges. One pivotal challenge was Isaac Newton's theory of planetary motion, as described in his *Philosophiae Naturalis Principia Mathematica* (Mathematical

Principles of Natural Philosophy), published in 1687. Inspired by what he perceived as an atheism inherent in Descartes' mechanical philosophy (which Descartes vigorously and sincerely denied), Newton conceived of a universe in which God had room to act. Newton's theory was widely interpreted as describing a universe almost totally devoid of matter, with objects embedded in vast reaches of empty space that could exert influences over each other from great distances by way of the force of gravity, interpreted by many as action at a distance between bodies separated by (sometimes enormous) distance. This was a radical departure from the mechanical view of the universe, which had envisioned a material but undetectable fluid called an *ether,* which completely filled in spaces between objects in the universe. The concept of an ether allowed distant objects to affect each other according to mechanical principles: as things moved through the ether they could form vortices and currents, which could in turn affect other objects. The presence of the ether allowed the mechanists to avoid the need for mysterious concepts such as forces. But Newton provided no explicit explanation for gravity—where it came from and why it existed—other than by arguing that it was God's will imposed on the universe. However, he used the concept of gravity to predict planetary motion with great mathematical precision, and it was very difficult for opponents to deny its power as an explanatory concept.

Some of those who studied life thought perhaps there might be a force analogous to gravity but unique to living things, which could explain life and generation (Newton himself in fact declared the same thing). The eighteenth century thus experienced a vigorous revival of vitalism, and many natural philosophers argued that there existed immaterial forces, unique to life, that explained generation. One major eighteenth-century example was *epigenesis,* the theory that living matter contained a vital principle that allowed it to self-organize. Advocates of epigenesis believed that, rather than being preformed, life acquired form through a process that was unique to living things. Epigenesis implied some sort of active organizing principle unique to life, a view that, again, had its origins in ancient Greek vitalism, and especially Aristotelianism.

A variety of observations seemed to support epigenesis. In 1741, Abraham Tremblay described his observation of the regeneration of an animal form. Tremblay was in fact the cousin of Charles Bonnett, who a year earlier, in 1740, had described his observation of parthenogenesis in the female louse. Tremblay observed regeneration in a simple animal called a polyp (now called a hydra): he cut the polyp in half, and both halves regenerated into full organisms (Figure 2.5). Tremblay performed additional experiments and found that, no matter how he cut up the organism, each piece would regenerate a complete organism. Tremblay's experiments clearly challenged preformation. If each organism was fully formed one within another in eggs, how then did new organisms get made from halves of an old organism?

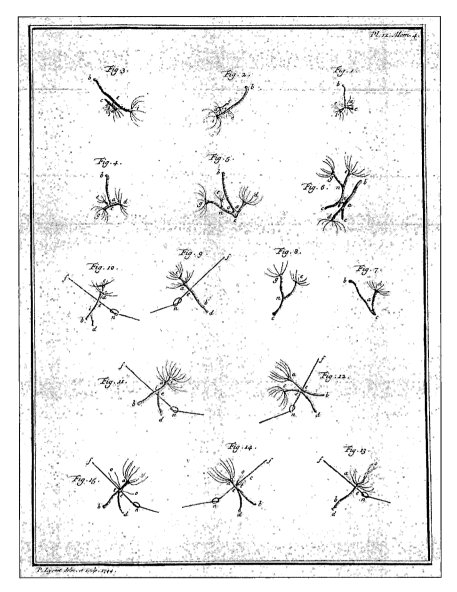

Figure 2.5: Abraham Tremblay's experiments with polyps (hydra). Image copyright History of Science Collections, University of Oklahoma Libraries.

Additional observations in the eighteenth century supported epigenesis. Caspar Friedrich Wolff described, in his 1759 *Theory of Generation,* his view of the epigenetic development of a chick embryo, and in particular the formation of a neural tube, an early stage in the development of embryos. Other important organs, especially parts of the nervous system, are formed from the embryonic neural tube. In his *Theory of Generation,* Wolff, inspired by Newton, outlined the concept of an essential force to explain the process of neural tube formation.

In his 1749 *Histoire Naturelle* (Natural History), a massive, multivolume publication, Georges-Louis LeClerc, Comte de Buffon outlined what he viewed as a compromise between the views of preformation and epigenesis (Figure 2.6). Buffon was an influential man, head of the famous Jardin de Roi (Garden of the King) in France, which contained a massive specimen collection from around the world. Buffon argued that life consisted of unique, indivisible material called *molécules organiques* (organic molecules), which were given shape by a type of force called *moules intérieurs* (interior molds). Buffon was inspired by Newton, and also by physical molds popular in the eighteenth century that were used to form sculptures. Buffon's *moules intérieurs* penetrated the body everywhere, so matter could, in effect, "pour" into them and produce all of the body's internal and external structures.

By the late eighteenth century, vitalism was dominant over mechanism in studies of life's processes. In this context, in late-eighteenth-century Germany, an influential movement called *naturphilosophie* (philosophy of nature) became very popular, influenced by several German vital-

Figure 2.6: Georges-Louis LeClerc, Comte de Buffon. Courtesy of the Library of Congress Prints and Photographs Division.

ist thinkers, in particular Johann Friedrich Blumenbach and Immanuel Kant, who together created the concept of the *bildungstrieb* (formative force), which was conceived as a life force of unknown cause but with clear measurable effects. The nature philosophers expanded vitalism to the whole universe, arguing that all of nature was like a living organism guided by mysterious forces. As described later, vitalism influenced a generation of German researchers to examine in detail early embryonic development.

CONCLUSION

Aristotle's influence on natural philosophy slowly began to wane in the late medieval period. From the fourteenth to the seventeenth centuries, Western Europe expanded, trade and wealth increased, new technologies such as the printing press enabled the mass reproduction of books, new societies were formed, and the known world expanded as a consequence of colonialism, microscopic studies, and other developments. In this context, Scholasticism came under increasingly aggressive challenges, and the mechanical philosophy in particular sought to replace Aristotelianism with a new, comprehensive theory that conceived of the universe as a vast machine, subject only to simple mechanical laws of matter in motion.

But living things, and especially the process of animal generation, continued to be a source of mystery. Using preformation theory, the mechanists attempted to elude the problem of the acquisition of form during heredity and development. But this was a mixed success, and the mechanical philosophy experienced increasing challenges by the late seventeenth and eighteenth centuries as vitalism experienced a resurgence in popularity. Isaac Newton's introduction of the concept of the force of gravity, with its unknown causes but very accurately measurable effects, inspired new explanations for life's processes that incorporated vitalist explanations of unknown forces, unique to living things. As described in the next chapter, vitalism's resurgence in the eighteenth century encouraged new explanations for generation, which were combined with new theories of the evolution of living things. While mechanism continued to be influential in some branches of biology, such as anatomy and physiology, vitalism was the dominant framework for these new conceptions of generation and evolution.

3

HEREDITY IN THE
NINETEENTH CENTURY

INTRODUCTION

In the preceding chapter we discussed preformation and epigenesis, two theories of generation offered, respectively, by mechanists in the seventeenth century and vitalists in the eighteenth century. As was noted, by the eighteenth century vitalism was experiencing a resurgence, while strict mechanist accounts of the nature of the universe and of life experienced increased criticism.

The successes of vitalism led to a renewed enthusiasm for understanding and theorizing about the origins, growth, and development of living things. Connected to this renewal was a gradual exploration of the possibility of *evolution*, the belief that forms of life might have been very different in the past from the way they were observed to be in the present. Theories of evolution were intimately entwined with theories of generation. The vitalist acceptance of a force responsible for acquiring form led quite easily to a belief that this force could allow forms to change over time.

This chapter discusses evolution and cell theory, two pivotal topics in the history of biology that led to major developments in theories of inheritance. The chapter begins with the development of evolutionary theories in the late eighteenth and nineteenth centuries, culminating with Charles Darwin's theory of natural selection; this section illustrates that evolutionary theories and inheritance theories developed in parallel, as biologists became increasingly interested in how organisms related to each other historically, across multiple generations. The chapter concludes by describing embryology and cell theory, crucial areas of research that originated primarily in Germany, inspired by vitalist views of the origin and development of life. This description of cell theory culminates with the development of the theory of *hard inheritance,* which argued against the prevailing notion of generation that connected the processes of heredity and development; as we will see, the separation of these processes was crucial for the development of twentieth-century genetics.

EARLY CONCEPTIONS OF EVOLUTION

Buffon, an influential eighteenth-century vitalist who theorized that animal generation was the result of *molecules organiques* given form by *moules intérieurs* (see Chapter 2), was also one of the first natural philosophers to consider the possibility of evolution. In a 1779 book titled *Epoques de la Nature* (Epochs of Nature), Buffon argued that the earth had gone through a series of changing stages, or epochs, which he suggested might have lasted for many thousands of years. In each epoch, Buffon argued, new *moules intérieurs* could arise that were best suited to a given epoch. As a result, Buffon believed that the forms of animals were different in different epochs; they evolved according to the nature of the given epoch's environmental conditions.

Buffon also believed that *moules intérieurs* could change slightly, in order to produce variations on a particular form of an animal. Buffon called this process *degeneration,* and he believed that this sort of process was responsible for creating sets of related animals, such as, for example, different sorts of wild cats—lions, tigers, cougars, and so on. Unlike many of his contemporaries, Buffon thus saw the living world as dynamic rather than static, experiencing a great deal of change over time. He argued that his concept of changing molds, combined with his theory of degeneration, provided a useful theory to explain this historical, changing dimension of the natural world.

Buffon's theory of a dynamic earth and his belief that life changes over time influenced a growth in interest in evolution. For example, Erasmus Darwin, grandfather of Charles Darwin (discussed later), speculated in his book *Zoonomia* (1794) that all living things might have evolved from a single common ancestor, an important concept similar to that which appears in his grandson's theory, almost a century later. Theories about evolution in fact became very popular for the vitalists, because it was assumed that the immaterial "life force," however conceived, would be flexible enough to allow for life to adapt to changing environmental conditions as time progressed. Unlike the mechanists, who avoided theorizing about generation and the acquisition of form, the vitalists felt free to speculate on the nature of generation and on how forms might have changed over time.

In this context, views that life had undergone changes in concert with changes in the earth's formation were commonplace. The advocates of *naturphilosophie* saw the universe as growing and changing like an organism, thereby explicitly equating evolution with organismal generation. The popularity of evolutionary theory also grew as increasingly detailed theories of the age of the earth emerged from the field of geology, suggesting that the earth was much older than even Buffon had imagined. Ideas about evolution were thus given a much longer timeline, and sophisticated theories came to be developed.

LAMARCK'S THEORY OF EVOLUTION

In the early nineteenth century, Jean-Baptiste de Lamarck described a very elaborate theory of evolution. Lamarck was a well-known French natural philosopher. In 1793 he was appointed to the Museum of Natural History in Paris (previously called the Jardin de Roi, for which Buffon had once been a curator). Lamarck was Professor of "Insects, worms, and microscopic animals," for which Lamarck coined the term *invertebrate*. Lamarck also popularized the term *biology* to refer to the study of living things. Lamarck began describing his theories on evolution in 1801, and, in an 1809 book titled *Philosophie Zoologique* (Zoological Philosophy), he summarized his theory.

There are two central components to Lamarck's full system of evolution. The first is his belief in what has been called an *inherent tendency toward complexity* (Figure 3.1). Lamarck argued that life has increased in complexity over time, from less complex to more complex organisms, with humans as the most complex. His theory was based on an ancient concept, the Great Chain of Being, which originated with ancient Greek philosophers and was widely accepted by natural philosophers. The Great Chain of Being was a hierarchical ordering of life in terms of complexity, with humans, assumed to be the most complex, at the top of the chain. Lamarck argued that simpler organisms could progress up the Great Chain of Being, evolving from simpler to more complex organisms.

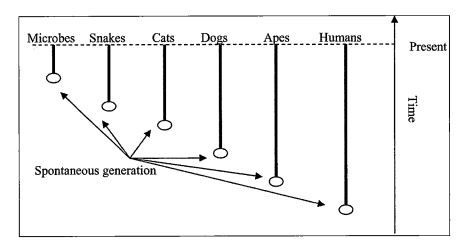

Figure 3.1: Jean-Baptiste de Lamarck's inherent tendency toward complexity.

Life began, Lamarck argued, with the *spontaneous generation* of the simplest organisms. Spontaneous generation was in fact a long-held theory, predating Lamarck, that states that life originates from nonlife. A lot of evidence suggested that this was the case: molds grow on rotting food, maggots seem to grow out of nowhere, and so on. Leeuwenhoek's observations of animalcules in the seminal fluid (see Chapter 2), for example, were the focus of many seventeenth- and early-eighteenth-century debates about spontaneous generation. Some thought that Leeuwenhoek's animalcules were spontaneously generated little animals, while others invoked putrefaction to explain their presence. By the early nineteenth century, spontaneous generation was the subject of significant debate and criticism, but it was still popular with mechanists.

Lamarck argued that these simplest of organisms then gradually evolved, increasing in complexity over time from very simple life forms to human beings, while passing through all the intermediate forms of complexity in the process. Furthermore, Lamarck argued, spontaneous generation was occurring all the time. This explains why, at any given point in time, there were a great variety of living forms varying in complexity.

The idea of an inherent tendency toward perfection sounds vitalistic: Lamarck seems to be implying that life has some sort of mysterious, internal force that causes it to progress. But, in fact, Lamarck explained this increase in complexity mechanistically. He argued that fluids carved channels in organisms over time, and these channels carried blood and other materials crucial for life to function. As these channels became more numerous, life became more complex.

The second component of Lamarck's theory is *the inheritance of acquired characteristics.* This is Lamarck's most famous, although often misunderstood, theory. Lamarck believed that organisms can be modified by their environment and that this modification can be passed on to their offspring. For example, if a blacksmith were to use his right arm more than his left, then his children would be born with a right arm stronger than the left. The theory is now discredited; acquired characteristics cannot be inherited.

Lamarck believed that the inheritance of acquired characteristics explains why life is not observed to be a linear arrangement of living things in a simple order of increasing complexity, which is what one would predict according his explanation of an inherent tendency toward complexity. The inheritance of acquired characteristics, according to Lamarck, explains why there exist variations on any given form; for example, it explains why there are so many different types of cats, or dogs, or fish. The inheritance of acquired characteristics was a theory of inheritance, not a theory of evolution; it was used to explain the great diversity of animals that are similar in appearance and the apparent gaps in what we would expect to be the orderly and predictable progression of living things that would be predicted by his theory of evolution via the inherent tendency toward complexity.

Lamarck's theory did not catch on until some time after his death, when, as we will see later, Darwin published his theory of evolution. But in Lamarck's lifetime his theory was attacked vigorously by Georges Cuvier (Figure 3.2), a strong opponent of evolution. Like Lamarck, Cuvier was a professor at the Museum of Natural History, and its most respected zoologist. Cuvier studied the comparative anatomy of animals; he compared living animals to each other and to fossils and saw no evidence of evolution in his studies. Cuvier found fossils that bore no resemblance to existing animals and argued that these fossilized animals were now extinct. In contrast, Lamarck did not believe that organisms went extinct, and he explained these observations by arguing that the animals had evolved into new forms.

Cuvier advocated a theory that he called the *correlation of parts*. He viewed animals as functional wholes. According to Cuvier, any change in an organism would disrupt the natural correlation and harmony of all

Figure 3.2: Georges Cuvier. Courtesy of the Library of Congress.

its parts. Cuvier's remarkably detailed anatomical studies seemed to support his theory. Cuvier was also suspicious of Lamarck's idea of the inherent tendency toward complexity, which he viewed as too mysterious and too much like vitalist explanations, now out of favor in France and England (but not, as we will see, in Germany). Cuvier was very influential and very wealthy, unlike Lamarck, and was consequently able to exert a significant amount of power in his battle with Lamarck. Their bitter dispute left Lamarck's reputation destroyed.

For much of the nineteenth century, evolutionists were in the minority. The influence of Cuvier and others had established the view that organisms were so well adapted to their existence that it was difficult to conceive how their anatomical structure could change. Mechanism also made a comeback in the nineteenth century, and for most nineteenth-century biologists, the role of the biologist was simply to study how animal functions worked mechanically, rather than to speculate on the relationship of form to function, a task that was once again was left to theologians. The nineteenth century in fact saw a great revival in popularity of Natural Theology, the view that the harmony of nature was evidence for God's existence and beneficence. In 1802, William Paley

Figure 3.3: William Paley. Courtesy of the Library of Congress Prints and Photographs Division.

(Figure 3.3) wrote *Natural Theology,* which was very similar to John Ray's 1691 *Wisdom of God Manifested in the Works of Creation,* discussed in the previous chapter. Paley's book was very popular reading in the nineteenth century, and it helped reinforce the influence of the mechanists. Paley outlined in detail what has been called the "divine watchmaker argument": if you were to take a walk deep in an uninhabited forest and come across a watch, surely you would not think it appeared there by accident, or through some sort of natural process. Its complexity would be evidence for you that someone had designed it. Similarly, said Paley, we should be surprised to find that complex, living things came about by some natural process. Rather, surely there is a "divine watchmaker," who crafted these complex living things on purpose.

Despite the renewed dominance of mechanism, some pockets of vitalism remained and produced important results. Researchers in Germany in particular, influenced by *naturphilosophie* (see Chapter 2), produced very important work in embryology, investigating how form was produced during the development of the early embryo. By the mid-nineteenth century, increased experimental evidence in German embryology was used as support for evolution, because of a belief that developmental stages imitated evolutionary stages. And, as described in more detail later, related research in embryology and cell theory revealed the existence of chromosomes and their role in inheritance.

British biologists in particular remained very resistant to evolution. Britain was very conservative politically at the time. The French Revolution was very violent and radical and was still a fresh memory for the British. Change was generally seen as a bad thing in England. Despite mounting evidence used to argue for evolution, British scientists often had counterexplanations. For example, Richard Owen argued that similarities in different organisms' anatomy was evidence not of evolutionary relationships but of God's use of one sound body plan with modifications for particular functioning needs. Louis Agassiz argued that vestigial organs—body organs that seemed to have once been functional but that were no longer used by an animal, such as the pelvic bones of aquatic mammals—were simply evidence that the Creator had a broader purpose when creating the animal's form; perhaps they were rudiments of organ development that would appear at a later time or were simply added by God for aesthetic reasons. British scientists were very suspicious of evolutionary theory, which seemed to them to be a radical attack on God and the State.

DARWIN'S THEORY
OF EVOLUTION BY NATURAL SELECTION

Into this context came Charles Darwin (Figure 3.4) with his theory of evolution by natural selection, described in his 1859 publication titled *On the Origin of Species by Means of Natural Selection; or, the Preservation of Favoured Races in the Struggle for Life,* commonly called *The Origin of Species* (Figure 3.5). According to natural selection, evolution was an open-ended product of random chance, rather than a predictable, directed process such as Lamarck's inherent tendency toward complexity. Darwin argued that some individuals in a population occasionally contain random variations in structure that happen to give them an advantage in their particular environment. Because these individuals have a slight survival advantage, they tend to outcompete less advantaged animals, and as a result they live longer and have offspring more frequently. As a result, they pass on their "evolutionary fitness" to their offspring, who themselves are more likely to survive and have offspring. Darwin argued that, over long periods of time, this process of "survival of the fittest" is sufficient to cause major changes in a group of organisms through a process called divergence: an original population with minor variations will, given enough time, evolve into separate populations with major differences (Figure 3.6).

THE ROLE OF HEREDITY
IN NATURAL SELECTION

According to Darwin, any given population of living things, such as a herd of cows, a flock of geese, or a forest full of oak trees, exhibits a

Figure 3.4: Charles Darwin. Image copyright History of Science Collections, University of Oklahoma Libraries.

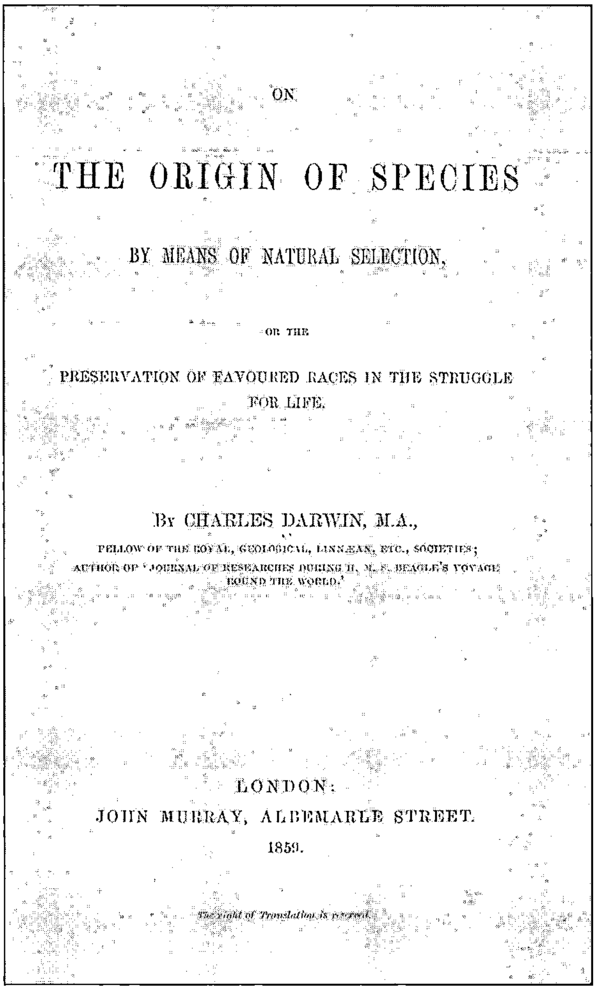

ON

THE ORIGIN OF SPECIES

BY MEANS OF NATURAL SELECTION,

OR THE

PRESERVATION OF FAVOURED RACES IN THE STRUGGLE FOR LIFE.

By CHARLES DARWIN, M.A.,

FELLOW OF THE ROYAL, GEOLOGICAL, LINNÆAN, ETC., SOCIETIES;
AUTHOR OF 'JOURNAL OF RESEARCHES DURING H. M. S. BEAGLE'S VOYAGE ROUND THE WORLD.'

LONDON:
JOHN MURRAY, ALBEMARLE STREET.
1859.

The right of Translation is reserved.

Figure 3.5: Title page of *The Origin of Species.* Image copyright History of Science Collections, University of Oklahoma Libraries.

great deal of variation among its individual members. This is an absolute requirement for natural selection to work: natural selection needs variation upon which to act. The first several chapters of the *Origin of Species* discuss variation almost exclusively. Chapter 1 is titled "Variation under Domestication," and in it Darwin discusses how breeders artificially select preferred characteristics from domestic populations of animals that show lots of variation. Chapter 2 is titled "Variation under Nature," where Darwin argues by analogy that a similar selection occurs in nature. Darwin then uses this observation of variation in populations to argue for natural selection: some variations are more advantageous for survival than others, and, because there is a struggle for existence, any such slight advantage will be important in this struggle. A population thus changes over time, by very gradual accumulation of small but favorable variations over very large timescales.

But from where did all of this variation in a population come? Why should such variation exist? In Chapter 5, titled "Laws of Variation," Darwin described a process he called *pangenesis,* in order to explain the origin of variation. According to pangenesis, every part of the body produces "gemmules," tiny bits of material that can grow like seeds into body parts. These gemmules, once produced, move to an organism's reproductive organs. During reproduction, gemmules from both parents mix together to produce offspring. The reason so much variation exists in a population, according to Darwin, is that gemmules can be affected and changed by the environment of the parents. These changes can then be inherited by offspring. In other words, and contrary to popular belief, Darwin believed in the inheritance of acquired characteristics, just like Lamarck did.

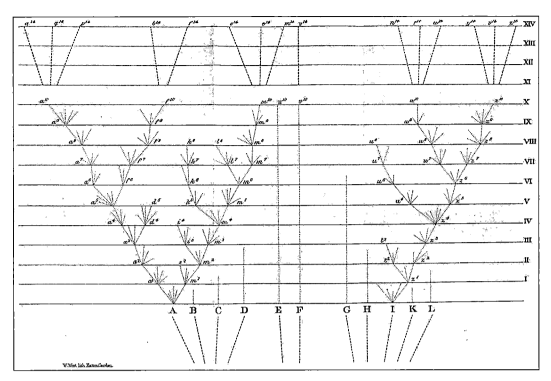

Figure 3.6: Darwin's drawing of divergence, from *The Origin of Species*. Image copyright History of Science Collections, University of Oklahoma Libraries.

DARWIN, LAMARCK, AND THE INHERITANCE OF ACQUIRED CHARACTERISTICS

One of the greatest myths in the history of science is that Darwin and Lamarck had very different views about the inheritance of acquired characteristics. According to this myth, Lamarck attempted to explain evolution by way of the inheritance of acquired characteristics, but Darwin's theory of natural selection refuted Lamarck's theory. The classic example is the description about how the giraffe evolved its long neck: according to countless textbooks, Lamarck believed that giraffes stretched their necks to reach leaves in tall trees; they then passed on this acquired "stretched neck" characteristic to their offspring. Darwin, in contrast, is depicted as arguing that some giraffes already had slightly longer necks, and consequently they outcompeted short-necked giraffes, survived more often, and passed on their longer necks to their offspring. On the surface, this seems like a reasonable dichotomy, since the theory of natural selection seems to preclude the need for the inheritance of acquired characteristics. According to natural selection, there were *already* giraffes with slightly longer necks, so there is no need to imagine that giraffes had to acquire this trait before passing it on. But when we read the *Origin of Species* we discover that, according to his theory of pangenesis, Darwin *also* believed in the inheritance of acquired characteristics. How to explain this confusion? Why has the myth of Darwin versus Lamarck existed for so long? Why did Darwin believe in both natural selection *and* the inheritance of acquired characteristics?

The crucial point is that the inheritance of acquired characteristics explains not evolution but the *inheritance of variation*. It is a theory of inheritance, not a theory of evolution. Lamarck's theory of evolution was not the inheritance of acquired characteristics but the *inherent tendency toward complexity*. However, Lamarck recognized that the ideal result of the inherent tendency toward complexity, with a set of animals in increasing order of complexity, did not really exist; often many organisms seemed to be very similar in their level of complexity but looked slightly different from each other, for example lions and tigers. Lamarck explained this sort of variation using the notion of the inheritance of acquired characteristics. Darwin explained variation in exactly the same way but elaborated it with his theory of pangenesis.

Both Lamarck and Darwin accepted the inheritance of acquired characteristics because it seemed very plausible. As described earlier, natural philosophers tended to see heredity and development as part of one process, called *generation*, a theory used at least as early as Aristotle's time that stated that, as the body grows and develops into an adult, bits of matter are put aside and used for inheritance. Darwin's theory of pangenesis needs to be understood in the context of this belief in generation. While we now know it is not true, it was a reasonable assumption that the same mechanisms that let the body grow into that of an adult are involved in forming that body in the first place. It was not until the discovery of the chromosomes in the late nineteenth century that heredity and development were proposed to be independent processes, and not until the early twentieth century that the concepts of heredity and development were separated and the concept of the inheritance of acquired characteristics discarded (see Chapter 4).

THE RESPONSE TO THE *ORIGIN*

By 1865, evolution was no longer controversial among biologists; Darwin's book and its enormous popularity—all copies of the first edition sold in a single day, and numerous editions appeared with great frequency—convinced people of it. Additionally, Darwin's supporters were quite influential, and England was no longer as conservative as it had been earlier in the nineteenth century. But Darwin's theory about how evolution occurred, by natural selection, in fact remained very controversial, and quite unpopular. It was a random, open-ended system, with no inclusion of the notion of *progress* in evolution. According to natural selection, humanity is simply a chance occurrence. By this time there was in England a strong belief in *social progress,* the use of science and industry to make society better, and the idea of *biological progress* was a natural extension of this belief. Biological progress was not clear in Darwin's theory, but it was very clear in Lamarck's notion of the inherent tendency toward complexity. Lamarck's theory of evolution gained a new and receptive following in this context, while Darwinists who advocated natural selection found themselves in the minority until the 1930s.

CELL THEORY, EMBRYOLOGY, AND HEREDITY

At about the same time that Darwin developed his theory of natural selection, a new line of research began to shed further light on the processes of heredity. This research culminated in the *cell theory,* a view of the cell as the fundamental unit of life for all organisms, and a description of the cell's involvement in both heredity and development. Some earlier research had preceded this nineteenth-century work; for example, the seventeenth-century microscopists first described cells. Robert Hooke discovered cells and described them in a celebrated book titled *Micrographia* (1665); in 1675–1679, Marcello Malpighi described what he called utricles, which were basic cellular structures of plants; and Nehemiah Grew described plant cells in some detail in *The Anatomy of Plants* (1682). But these early observations and their interpretations were debated, and many dismissed them as unimportant, or as transient, or sometimes as simple artifacts of microscopic observation. It was not until the late nineteenth century that analyses of cells really became a comprehensive field of research; with this research came an increased focus on the nucleus as the cell's hereditary material.

In 1831, Robert Brown, a Scottish botanist, concluded that the dark, solid-looking thing many people had at that point observed in cells was not an artifact but instead an actual component of cells. Brown named it the nucleus and observed it in many different types of cells. While Brown discovered the nucleus, the German biologists Jakob Mathias Schleiden and Theodor Schwann did collaborative research and performed various microscopic observations over the next decade (Figure 3.7), which led to their articulation, in 1839, of the *Cell Theory.* Schleiden and Scwann saw the nucleus as the organizing center of cellular growth and development. They argued that a nucleus forms from spontaneous organization of granular substances inside or (for Schwann) outside an existing cell. Once this occurs, a membrane surrounds the new nucleus, and it becomes a new cell. In the process, they argued that cells were the basic, fundamental units of life. They argued that cells were the simplest living things and that aggregates of cells make up more complex forms of life.

A variety of research in the 1850s demonstrated that cells divided to produce other cells. In 1858 Rudolf Virchow wrote *Die Cellularpathologie* (Cellular

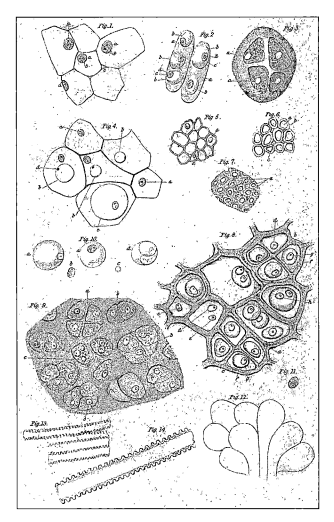

Figure 3.7: Theodor Schwann's drawings of cells. Courtesy of the American Philosophical Society.

Pathology), which summarized the experimental work of Virchow and others and argued that the cell was a living, replicating unit that cannot arise from nonliving material—an explicitly vitalist account of the origins of life. Virchow coined the phrase *omnis cellula e cullula,* all cells from cells, which became a widely accepted view of the origin and inheritance of living things.

By the 1860s, a different theory of cells was more accepted, one that placed less emphasis on the role of the nucleus. From 1835 to 1860, a variety of French and German microscopists had observed a viscous, jelly-like fluid that filled the interior of all cells. In an 1861 paper, Max Schultze, a German cell theorist, used these observations to outline what he called the "Protoplasm Theory." Schultze argued that the protoplasm, not the nucleus, was the essence of the cell and the fundamental unit of life. In 1868, Thomas Henry Huxley, a British botanist, wrote a paper called "On the Physical Basis of Life," which popularized the Protoplasm Theory and made protoplasm a household word.

But the Protoplasm Theory enjoyed only a brief period of dominance; the 1870s saw an increased focus on the nucleus during cell division and on the early stages of development. It had been known since 1850 that, during cell division, the nucleus dissolves, nuclear material moves to the center of the cell, and two new nuclei are formed. The two new nuclei are then separated by a membrane, resulting in the formation of two new cells. In the 1870s, several researchers elaborated on nuclear division, describing the nuclear material as looking like bundles.

Other researchers interested in embryology studied the fusion of sperm and egg cells and speculated that their nuclei somehow come together during fertilization in order to produce an embryo. In 1883, the Belgian Edouard van Beneden discovered that these nuclei remain as distinct material in the fertilized egg and separate again during subsequent cell division. Van Beneden also realized that the two nuclei in fact contain only half the material of the full nucleus and that the two together contain the full nuclear material of a normal cell.

In 1887–1888, this process, called *reduction division* (now called *meiosis*), was studied and confirmed by the Germans Theodor Boveri and August Weismann, who found that, during division of the cells that form sperm and egg, the nuclear material divides once while the cell divides twice. This is in contrast to the normal type of cell division (today called *mitosis*) that produces the remainder of the cells of the body, in which both the nuclear material and the cell each divide twice. The sperm and egg cells, as a result of reduction division, have only half the original nuclear material, unlike other cells in the body, which receive a full set of nuclear material from their parent cells following mitosis (Figure 3.8). Weismann, who was interested in heredity, argued that reduction division is required in order to prevent the doubling of the nuclear material in every successive generation. The process was confirmed in 1890 by Oscar Hertwig, who suggested that these bundles of nuclear material, given the name *chromosome* in 1888 by Heinrich Waldeyer, are responsible for contributing the male and female hereditary characteristics. Theodor Boveri built on this work and performed detailed experiments from 1887 to 1909 that showed that normal development requires a specific number of chromosomes from the male and from the female.

Also in the mid- to late 1880s, another line of embryological research was examining the relationship between whole cells and development by focusing on the whole cell and the whole organism rather than on the nucleus. In an 1890 publication called *Die Entwickelungsmechanik der Organismen* (The Developmental Mechanics of Organisms), Wilhelm Roux described a series of experiments in which he killed one of the two cells of the first cell division of a fertilized frog's egg. The remaining egg grew into a half embryo. Roux concluded that each of these two cells held only half the full set of chromosomes. He theorized that only a portion of the full set of chromosomes gets

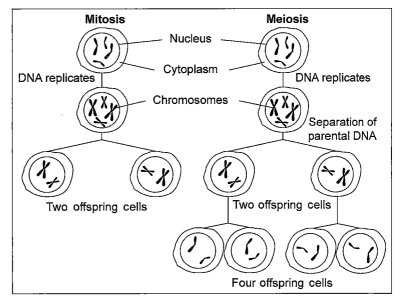

Figure 3.8: Meiosis and mitosis. Illustration by Jeff Dixon.

transmitted during cell division. Thus, Roux believed that an embryo's development involved the replication of one initial cell and that the subsequent formation of specialized body parts was a result of cells containing only portions of chromosomal material that encoded for that specialized part.

We now know that this conclusion was incorrect: every cell in our body contains a full set of chromosomes. Each cell becomes specialized not because, as Roux thought, it has only a portion of the chromosomes but because only a portion of the chromosomes are actually used by the cell. But Roux's incorrect conclusion stimulated a theory that is central to our modern conception of genetics. An important question arising from Roux's theory was the following: How did the full chromosome set get passed on to the next generation if only a portion of the chromosomes are passed on to new cells during development? In 1892, August Weismann wrote *Das Keimplasm* (The Germplasm) to answer this question, and in the process he outlined something close to our modern theory of heredity.

Weismann argued that there are two sets of cells, the germ cells (or germplasm) and the somatic cells (or somatoplasm). Weismann accepted Roux's theory that somatic cells contain only partial sets of chromosomes that contain instructions for the body part that the cell will ultimately become. But Weismann argued that the germ cells, which he said were responsible for heredity, contain the full set of chromosomes and pass these on to future generations. The germ cells remain intact during the stages of development from a fertilized egg to a formed embryo, through growth to an adult and then pass on the full chromosomal set, unchanged, to offspring. The combined ideas of Roux and Weismann constitute the Roux-Weismann Cell Theory of Development: cell division causes development, and chromosomes contain the information needed for development; somatic cells have only a partial chromosome set and so differentiate into various body parts; and germ cells pass on a full chromosome set during heredity.

In 1883, Weismann wrote *Uber die Vererbung* (On Heredity), which summarizes how his research on germ cells and somatic cells supported a theory of what was later called *hard inheritance*. Weisman argued that only the germ cells are involved in heredity, while somatic cells are involved only in the development of an individual organism. The germ cells, he argued, are not affected in any way by the somatic cells; therefore, any changes in the body of an organism do not alter the hereditary material. Weismann thus left no room for the Lamarckian concept of the inheritance of acquired characteristics, which almost everyone at this point followed to some degree, including, as we have seen, Charles Darwin himself.

Weismann thus argued for continuity of the germplasm and separated heredity from development. The theory marks an important turning point for studies of heredity. As discussed earlier, up to this point biologists believed that heredity and development were two components of the same process—*generation*—and that the hereditary material was simply leftover material from development that was used to pass on information about the body to the offspring. Generation connected a parent to an offspring but supplied no connection—no *continuity*—across multiple generations of ancestors. Weismann argued, in contrast, for the *continuity of the germplasm*—the hereditary material, according to Weismann, was transmitted unchanged across many generations of organisms.

Since almost everyone believed in the inheritance of acquired characteristics, Weismann's theory was widely dismissed. Many saw hard inheritance as a revived sort of preformationism, because, like preformationism, it required the belief that some sort of form had existed throughout the entire ancestral history of an organism. Weismann thus had many opponents, but he himself became a great opponent of the inheritance of acquired characteristics: in the late 1880s, he cut off the tails of nine hundred young white mice in 19 successive generations and demonstrated that each new generation was born with a full-length tail. The final generation, he reported, had tails as long as those originally measured on the first and exhibited no inheritance of their parents' "acquired" lack of a tail. But this experiment was widely dismissed: biologists held a much more complex view of how acquired characteristics are inherited. Darwin's pangenesis, for example, could involve gemmule migration earlier in development than when Weismann did his experiment. Weismann was influential in the movement away from theories about the inheritance of acquired characteristics only after the turn of the century, with the discovery of a paper written in 1865 by an obscure German monk named Gregor Mendel, to whom we turn in the next chapter.

CONCLUSION

The resurgence of vitalism in the eighteenth century led to conceptions of evolution, the view that organisms could change in form over time. Buffon's early theories of evolving interior moulds were followed by other more elaborate theories, notably Lamarck's inherent tendency toward complexity and Darwin's theory of natural selection. Both of these theories focused primarily on evolutionary explanations, but crucial to both Darwin and Lamarck was the need to explain variation in animal forms that could easily be observed. For Lamarck, variation was a complicating factor in his theory of evolution, and it was explained by his theory of an inherent tendency toward complexity. For Darwin, variation was a necessary precursor to the possibility of evolution by natural selection. To explain such variation, both relied on earlier conceptions of generation that assumed that heredity was related to individual development. Lamarck's theory of the inheritance of acquired characteristics and Darwin's theory of pangenesis were very similar to these earlier conceptions and to each other: each assumed that, during development, material from an organism's various body parts migrated to the germ cells to take part in reproducing the next generation. Such theories of generation allowed for the inheritance of acquired characteristics and linked offspring to their parents but not to earlier generations.

The growth of embryology and cell theory in the late nineteenth century culminated in a redefinition of inheritance as a process completely independent of development. The discovery of the chromosomes and the observation of the process of reduction division led to August Weismann's theory of hard inheritance: Weismann's *somatic cells* played no role in inheritance but instead were involved only in individual development. A separate set of what Weismann called *germ cells* passed on chromosomal hereditary material to the next generation. These germ cells were independent of the rest of the body's cells and were unaffected by an individual's acquired characteristics. These cells harbored the hereditary material, keeping it intact and preserving its integrity so that it could be passed on, unchanged, from generation to generation.

Cell theory encouraged a growing interest in heredity as an independent object of study. New questions arose: how did the newly discovered chromosomes contain all of the instructions necessary to make a new organism? What was the nature of the hereditary material? As we will see in the following two chapters, these questions and more led to a great period of debate among those interested in heredity. This debate did not get fully resolved—by the so-called Modern Synthesis—until well into the twentieth century, and this resolution was a necessary precursor to a conception of the gene as a physical unit of hereditary information.

4

GREGOR MENDEL AND THE CONCEPT OF THE GENE

INTRODUCTION

Both evolution and theories of generation were fundamentally altered by the twentieth-century developments in heredity known as *Mendelism*, named after Gregor Mendel (Figure 4.1) and eventually described using the modern term *genetics*. Mendelism refers to the discovery that inheritance is under the control of individual units of hereditary information, now called genes. A primary result of Mendelism is that inheritance patterns and frequencies can be predicted using controlled breeding studies that calculate the proportions of offspring with alternative forms of a trait controlled by a given gene. Close historical studies in the past 15 or 20 years have revealed some surprises regarding the connection between Mendel's nineteenth-century work and the rise of genetics in the early twentieth century. Most historians now believe that Mendel himself most likely was not studying heredity. As described later, Mendel likely was pursuing an earlier research agenda involving attempts to determine the nature of species formation by interbreeding different varieties of plants in order to attempt the production of intermediate forms. But by the early twentieth century, Mendel's work was used to conceive of hereditary units that came to be called genes. This in turn led to speculation about where genes were located and about their composition.

The popularity of Mendelism also led to speculation about the best means to address many major social problems in early-twentieth-century society. Poverty, crime, alcoholism, and other social issues of the day increasingly became viewed as resulting from Mendelian inheritance. The simple yet powerful laws of Mendelism encouraged the belief that simple measures could be taken to

Figure 4.1: Gregor Mendel. National Library of Medicine / Photo Researchers, Inc.

dramatically improve the human race—typically by preventing "inferior" people from breeding. This belief was called eugenics, and it became a major social movement, armed with Mendel's laws of inheritance to justify coercive state measures to sterilize the genetically "unfit."

GREGOR MENDEL

Gregor Mendel was a monk of the Augustinian Monastery in Brünn, Austria (now Brno, in the Czech Republic). Mendel had a relatively modest education: he spent two years at the University of Vienna (1851–1853), where he was introduced to the new ideas of evolution, the emerging cell theory, research involving plant hybridization experiments, and, especially, mathematics. In 1865, Mendel published a paper in German called "Versuche über Pflanzen-Hybriden" (Experiments on Plant Hybrids), in which he described the results of his hybridization experiments using pea plants. He performed additional experiments on another plant from the genus *Hieracium* (Hawkweed) but did little science after he became the monastery's abbot, in 1868.

Mendel therefore had a very modest scientific career, yet today he is widely described as the father of modern genetics. How did this obscure scientific figure earn such a reputation? Mendel's 1865 paper attracted little excitement upon publication; even Darwin himself had a copy of Mendel's paper in his possession but had apparently given it no notice whatsoever. But, in 1900, the paper was "rediscovered," the term commonly used, by three men who saw in it support for their own argument that there exist distinct units of hereditary information that are transmitted from parent to offspring. But before describing this 1900 development, let us consider the contents of Mendel's 1865 paper.

Mendel studied seven characteristics of pea plants (Figure 4.2) from several *true-breeding* varieties—the particular variety always displayed the same form of a given trait of interest. Mendel bred tall plants with short plants, plants with different seed shapes (round or wrinkled) and colors (yellow or green), and plants with several other traits. In his breeding experiments, Mendel observed mathematical regularities that had not previously been documented. To understand these, consider Mendel's results for plant height.

When Mendel bred tall and short plants, all offspring were tall:

> *Experiment 1:* tall plants x short plants
> →tall offspring

The other characteristic, *shortness,* seemed to disappear. Mendel described the *tall* character as being *dominant* over the short, which he in turn called *recessive.*

In his second experiment, Mendel self-fertilized the offspring from his first experiment (plants have both male and female gametes and can therefore fertilize themselves). In this experiment, Mendel discovered a 3:1 ratio of tall to short plants— three-fourths of the resulting progeny were tall and one-fourth were short.

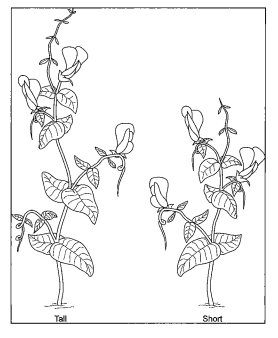

> *Experiment 2:* tall offspring self-fertilized
> →3:1 ratio of tall to short

Figure 4.2: Mendel's pea plants. Illustration by Jeff Dixon.

On the basis of this experiment, Mendel assumed that these tall offspring of Experiment 1 must still somehow contain the potential of their parental short varieties, despite their being tall. He called these plants *hybrids,* different from their tall true-breeding parents because they retained the potential for producing short offspring.

For his third experiment, Mendel self-fertilized the offspring of his second experiment. He found that the short plants were pure-breeding or true-breeding—all of their offspring were short. When he self-fertilized the "three-fourths tall" group, he found that one-third of this group were true-breeding, producing all tall offspring, but two-thirds of the group acted like their parental hybrids, producing the same 3:1 ratio that he had observed in experiment 2.

> *Experiment 3a:* short plants self-fertilized → short
>
> *Experiment 3b:* "3/4 tall" self-fertilized
>
> ↓
>
> 1/3 true-breeding tall (produced all tall offspring)
>
> 2/3 hybrid tall (produced 3:1 ratio of tall:short offspring)

Mendel thus concluded that the 3:1 ratio of tall to short plants, observed in experiment 2, was in fact more properly understood as a 1:2:1 ratio of tall to hybrid to short plants.

Experiment 2 reinterpreted:

tall offspring self-fertilized

↓

3 tall: 1 short

OR

1 true breeding tall: 2 hybrid tall: 1 true breeding short

Mendel got the same results for all seven characters he tested. He concluded that the first generation of offspring (which Mendel called the first filial generation, or F1) showed only the dominant character, but they retained what he called "the hereditary potential of both characters." This hereditary potential was transmitted to the second generation of offspring (F2). The two characters of a hybrid plant were transmitted independently of each other: sometimes a hybrid plant would transmit the short characteristic, and sometimes the tall. This process is known as *independent assortment*. Mendel realized that four possible combinations of characters could be transmitted to the offspring of two hybrid plants: the tall character from both plants (and so the offspring would be tall); the tall from the first plant and the short from the second, with the resulting offspring being tall because of dominance; the short from plant 1 and the tall from plant 2, with the offspring again being tall because of dominance; and, finally, the short character from both plants, with the offspring being short.

Mendel argued that the ratio he got represented a combination series—a standard mathematical term that Mendel knew well, having studied mathematics at Vienna. Using the symbols "A" for "tall" and "a" for "short," Mendel explained that his results conformed to what would be predicted based on all possible combinations of A and a:

$$A \times a \rightarrow A + 2Aa + a$$

This theory explains his observations. Mendel also did experiments studying multiple traits at once and used the same mathematical concepts to explain his results—that is, he argued that different traits are inherited independently of each other. When he studied two traits at once, for example, the results matched the expected ratio from combining two different combination series:

$$A + 2Aa + a$$
$$B + 2Bb + b$$

Interestingly, it is now known that not all of these traits that Mendel studied are transmitted independently of each other. Mendel should not have obtained the results he did. Additionally, Ronald Fisher, an early-twentieth-century statistician who studied genetics, concluded that even if the traits *were* all transmitted independently of each other, it was statistically very unlikely that Mendel's observations would be so close to the ratio predictions he had made. His results seem to be too good to be true, and to this day historians are unclear about the implications of this. But, regardless of this issue, for our purposes it is important to note that Mendel was mathematically inclined: he reasoned out this theory of transmittance using very simple and straightforward probability theory. Mendel was in fact very struck by the mathematical regularities of his results, and he suspected that they held significance.

MENDEL'S "REDISCOVERY"

Mendel's paper was found in 1900 by Carl Correns, Hugo de Vries, and Erich Tschermak, who at this point were interested in heredity and who were doing hybridization experiments to get some insight into its mechanism. It was then heralded as a landmark paper that was tragically ignored in Mendel's own lifetime until "rediscovered" by Correns, De Vries, and Tschermak. Mendel was given the status of founder of genetics, and, as described later, genetics gradually became a major new field of research.

Various ideas were proposed about why Mendel's work was neglected: Mendel was an obscure figure in science, not well connected to the rest of the scientific community, and Brunn was not a leading scientific center; he published in an obscure journal so his paper was not widely read; he was "ahead of his time," and it took some time before his brilliance was recognized; hereditary theory at the time was confused, and once it became clearer (especially once cell theory clarified the function of chromosomes), the importance of Mendel's work was recognized; finally, Mendel's one scientific contact, Carl von Nageli, an important German embryologist involved in cell theory work, encouraged Mendel to replicate his research in another plant, Hawkweed, which did not show the same ratios. It is now known that hawkweed is *apomictic;* it can reproduce asexually, which distorts the expected ratios. According to this story, Mendel, disappointed by the Hawkweed results, gave up on science.

For much of the history of biology, one or more of these explanations was accepted, and "Mendel's neglect" has become one of the most dramatic stories in the history of science. But the basic premise of these explanations, that Mendel was studying heredity, has in fact been increasingly challenged since the early 1980s (see sidebar). Many historians now believe that the traditional image of Mendel as an early scientist interested in heredity is in fact a myth, resulting from confusion about what Mendel was actually trying to do. According to this interpretation, Mendel was not, in fact, trying to discover the laws of inheritance; rather, he was testing whether hybridizing different plant varieties could produce new *intermediate forms.*

Was Mendel a "Mendelian"?

Many historians now believe that Mendel's research was most likely a test of the hybridization theory of species formation, first proposed by Carolus Linnaeus. Linnaeus, a Swedish natural historian, collected massive amounts of life specimens from colonized areas and created a very influential species classification system that is still used today, with some modifications. Linnaeus created the classification categories of Kingdom, Class, Order, Genus, Species, and Variety; he also created binomial nomenclature, the identification of organisms using a Latin two-name system corresponding to their genus and species, such as Homo sapiens for humans.

Buffon, a great opponent of Linnaeus, had argued that Linnaeus' classification system did not take into account the possibility that life can evolve over time. Linnaeus was opposed to this idea of evolution, but he allowed that perhaps new species could be created by hybridizing closely related plants; according to Linnaeus, God created the genus, and new species could perhaps arise by hybridization. Mendel, who may in fact have been an anti-evolutionist, was testing this theory. In fact, many contemporaries of Mendel bred plants for the same purpose. Interestingly, with his interpretation, Mendel's Hawkweed experiments gave better results than did his pea plants; their offspring had a form intermediate to their parents, suggesting that hybridization could produce new forms and, therefore, potentially new species.

In 1900, the scientists who discovered Mendel's paper had assumed that Mendel was trying to understand the laws of inheritance, because his description of characters, one from each parent, being transmitted independently and his ideas of dominance and recessiveness both clarified what was then known about heredity from cell studies and subsequent breeding experiments.

(continued)

WILLIAM BATESON, HUGO DE VRIES, AND THE RISE OF GENETICS

Gregor Mendel's work appealed to a number of twentieth-century researchers who studied heredity and evolution. One important person who greatly popularized Mendel's work was William Bateson. In the 1890s, Bateson was critical of the Lamarckian view of inheritance and focused on studying heredity distinct from development, although, interestingly, he was opposed to Weismann's work and the relevance of chromosome studies generally, believing instead that heredity somehow involved the whole cell.

Bateson believed in evolution but not in natural selection, a common view at this time. Natural selection suggested that nature selects the fittest organism from a population that exhibits a continuous set of variations. In contrast, Bateson argued that the evolution of a species must involve something more discrete than continuous, like a sudden mutation that quickly produces a new species. In his 1894 *Materials for the Study of Variation,* Bateson argued that "Species are discontinuous: May not the Variation by which Species are produced be discontinuous too?" Bateson thus created an alternative to Darwin's continuity thesis.

After Mendel's paper became well known, Bateson saw in Mendel's experiments evidence for his view that species formation involved dramatic changes in form. Mendel's paper did not demonstrate continuous variation in the characteristics that Mendel studied. To Bateson and others, this appeared to demonstrate that there existed hereditary units of information that determined whether a plant became, for example, either short or tall, with no intermediate options. This seemed to show that heredity was discontinuous rather than continuous.

Bateson was very influential, and his outspoken support of Mendelism greatly influenced its rise to popularity. In 1902, Bateson wrote *Mendel's Principles of Heredity: A Defence*, in which he coined the term *allelomorph* (later called *allele*), defined as alternative forms of any given trait. Height, for example, appeared to have short and tall allelomorphs, according to Mendel's experiments. Bateson also coined the terms *homozygote* to designate true-breeding organisms and *heterozygote* to designate hybrids. In 1905, Bateson coined the term *genetics* in a letter to Adam Sedgwick, and in 1906 he used the term at the Third International Congress on Hybridization and Plant Breeding. The name of this conference was changed to Third International Congress on Genetics. In 1910, he founded, with Reginald Punnett, the *Journal of Genetics*. From 1900 to 1915, Bateson was a leader in establishing genetics as a discipline; given his role, perhaps Bateson rather than Mendel is the more appropriate "Father of Genetics."

Several other researchers made similar contributions to genetics during this time period. Hugo de Vries, one of the discoverers of Mendel's paper in 1900, had in 1889 written *Intracellular Pangenesis*, in which he had argued that cells contain physical hereditary units called *pangenes*. De Vries' description of pangenes was quite similar to the later Mendelist view of hereditary units. Bateson greatly supported de Vries theory; they became friends, and de Vries in fact told Bateson about Mendel's paper in 1900. In 1901, de Vries published *Die Mutationstheorie* (translated into English in 1910), in which he argued, like Bateson, that evolution occurs by way of discontinuous evolution, involving major pangene mutations.

Another researcher important to early Mendelism was Wilhelm Johannsen. In 1909, Johannsen shortened de Vries' pangene to *gene*. Johannson also defined *genotype* as an organism's genetic makeup, in contrast to its *phenotype*, or physical appearance. According to this terminology, still used today, Mendel's tall hybrid plants were genotypically heterozygous, while phenotypically tall.

(continues)

But Mendel's idea of independent assortment came straight from mathematics training, while his idea of dominance appears to have come from his education in traditional Aristotelian Scholasticism. Aristotle used philosophical concepts of actuality and potentiality, upon which Mendel appears to have based his ideas of dominance and recessiveness. By the early twentieth century, new ideas about heredity made Mendel's results take on a significance they simply did not have in his own mind, or in the context of the plant breeding studies of his time.

CONTINUOUS VERSUS DISCONTINUOUS EVOLUTION

Bateson, de Vries, Johanssen, and others helped to acquire a secure scientific status for Mendelism, despite opposition from the Darwinists and biometricians. As briefly described earlier, the rise of Mendelism was accompanied by a growing division between the Darwinists and biometricians, who advocated a continuous theory of evolution, and the Mendelists, who believed that evolution resulted from discontinuous changes in form. Recall from Chapter 3 that Charles Darwin described his theory of Pangenesis in order to explain the presence of individual variation in a population, a necessary precursor for natural selection to function. Pangenesis involved the production of gemmules from different areas of the body, their migration to the germ cells, and their subsequent mixing together during reproduction to produce an offspring with features of both parents. Pangenesis was a *blending theory* of inheritance: gemmules were conceived as almost like a liquid, and reproduction would result in their mixing together to produce offspring with forms intermediate to its parents. Some of Darwin's critics, and in fact even many of his supporters, saw in this theory a critical flaw. Consider height: according to Pangenesis, a tall and a short plant will produce offspring of intermediate height. The offspring thus exhibit less variation than did their parents, and the next generation of offspring will exhibit even less variation. Eventually there would be no variation left in a population upon which natural selection could act. Each successive generation of organisms would show less variation than the previous, and eventually the entire population would look the same. Even new variations that might arise spontaneously (such as the unexpected appearance of a mutant plant) would quickly be diluted, much like dropping a drop of black paint into a bucket full of white paint.

Since Pangenesis could therefore not explain the presence of variation in a population, breeding experiments became a popular means to determine the nature of inheritance and its effect on variation in a population. A new generation of experimental breeders began to argue that blending inheritance did not hold up to experimental scrutiny: although simple observations of populations in nature might suggest that offspring always have forms intermediate to their parents (children, for example, tend to have a height intermediate to their parents), under controlled experimental conditions offspring often revert to their parental forms. It was in this context that Mendel's paper took on its new significance, as an elegant description of contemporary criticisms of the blending theory of inheritance and proof that evolution was probably discontinuous rather than continuous.

More than simply conceptual differences were at work in maintaining this division between supporters of discontinuous and those who accepted continuous theories of inheritance. The two groups had different objects of study: the Mendelists, like Mendel, studied small groups of organisms in controlled environments, while Darwinists studied large populations in nature. Mendelists were experimentalists, while Darwinists applied more abstract statistical models to explain evolution. There were also a lot of personal conflicts and bitter, inflammatory arguments between the two groups. Darwinists saw Mendelism as another weapon Bateson used to attack their statistical methods and so dismissed it as providing no insight into evolution. The modern synthesis of Darwinism and Mendelism—and therefore of evolution and genetics—had to wait until the 1930s, with the development of new mathematical concepts and the conception of the chromosome theory of inheritance, the subject of the next chapter.

MENDELISM AND EUGENICS

Mendelism was a key turning point in understandings of heredity, but its popularity and influence extended beyond scientific disciplines and into some of the most pressing social issues of the early twentieth century. The rise of Mendelism was accompanied by the rise of eugenics, an early-twentieth-century movement to breed better humans; the relationship between Mendelism and eugenics is explored in this section.

The term *eugenics* was coined in 1883 by its creator, Francis Galton (Figure 4.3), as "the study of the agencies under social control that may improve or impair the racial qualities of future generations, either physically or mentally." Eugenics involved encouraging the "better fit" to breed more and the "less fit" to breed less. Galton, a cousin and great admirer of Darwin, believed that all characteristics obeyed the laws of inheritance, including complex human mental and moral characteristics such as intelligence, criminality, poverty, sexuality, and much more. Galton advocated positive eugenics and negative eugenics to encourage improvement of the human race. Positive eugenics involved encouraging the "well-bred" to reproduce more. Various activities were held to popularize the positive eugenic goal, including state fairs, educational initiatives, and the so-called "fitter families contest" (Figure 4.4); the goal was to influence legislation that benefited the professional middle-class, thus providing an incentive for them to have large families. Negative eugenics involved discouraging breeding among

Figure 4.3: Francis Galton. Courtesy of http://galton.org.

the poor and among non-European people, who were assumed to be inferior to the white, upper-class society of which Galton and other eugenicists were members.

Galton and his student Karl Pearson are considered the founders of *biometrics,* the statistical analysis of biological populations. Together they invented a great many standard statistical techniques that are still in use today, most of which were designed for the purpose of studying eugenics in large populations.

While Galton and Pearson founded eugenics, the movement was also related to the rise of *social reform* in nineteenth-century Britain. Many late-nineteenth-century social groups were concerned about what was often described as the "problem of the rising urban poor." It was widely assumed by the vast majority of the upper class that poverty was a consequence of inferior capacity. Social-group members were concerned about things like alcoholism, moral degeneracy, crime, feeblemindedness, syphilis, sanitation, birth control, and physical and moral hygiene. These issues were often tied together with a general view that society was deteriorating because of the excessive breeding of the poor, while the upper classes were having smaller families.

Eugenics was very connected to the rise of Mendelism. While the movement originated in Britain with Galton, and with various social reform movements, British eugenics remained somewhat modest in its effects on government policy. British eugenics relied on Galton's and Pearson's statistical analyses, which were often complex and not very good at arousing the passions of those

Figure 4.4: Fitter Families contestants at a Georgia state fair. Courtesy of the American Philosophical Society.

unfamiliar with them. But, with the rediscovery of Mendel's laws, eugenics became influential in many countries. Eugenicists believed that all characters obeyed laws of Mendelian inheritance, including human mental characteristics, and that one could determine familial inheritance patterns for many complex mental and moral traits. The primary tool of Mendelian-based eugenics was not statistics but the *pedigree,* which could be used to provide a striking visual depiction of trait inheritance (Figure 4.5).

Eugenics became most popular in the United States, where commercial interests in animal breeding encouraged the popularity of Mendelism. In 1908, Charles Benedict Davenport formed the Eugenics Committee within the American Breeders Association, an influential group created to unite practical breeders with scientists studying inheritance in order to improve commercial breeds of plants and animals. In 1910, Davenport founded and became Director of the Station for the Experimental Study of Evolution (SEE), funded by the Carnegie Institution, a philanthropic organization founded in 1902 by Andrew Carnegie, a wealthy industrialist, to fund research for the improvement of mankind. Also in 1910, Davenport founded the Eugenics Record Office (ERO) at Cold Spring Harbor on Long Island, which became a central place for eugenics meetings, research, and data collection and for the creation and distribution of popular advertisements about eugenics. The ERO also trained fieldworkers to travel and collect hereditary information about families considered unfit and used pedigrees to illustrate Mendelian inheritance in humans. The ERO

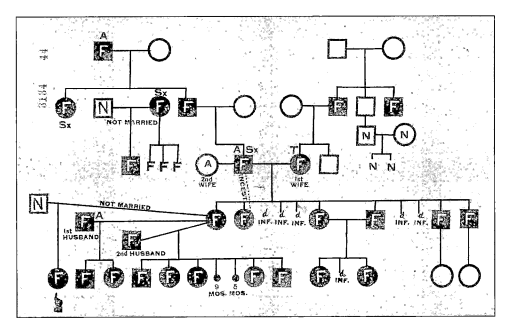

Figure 4.5: Pedigree of Carrie Buck, diagnosed as "feebleminded." Courtesy of the American Philosophical Society.

lasted until 1939, and, when it closed, the eugenics movement more or less ended as well.

Davenport and the ERO greatly popularized negative eugenics policies by state governments. Many policies were created that restricted immigration from areas considered to be inhabited by genetically inferior people, especially Eastern Europe and Africa. More alarmingly, laws were implemented that allowed for the sterilization of those assumed to be "genetically unfit." In 1927, the U.S. Supreme Court upheld a Virginia statute that instituted compulsory sterilization of those considered to be feebleminded. The case, *Buck v. Bell,* resulted in the sterilization of Carrie Buck (Figures 4.5, 4.6), a Virginia woman who had been committed to a mental institution primarily for acts of "immorality" (in fact she had been raped by her cousin). The case legitimized and upheld the constitutionality of compulsory sterilization; by the 1920s, about 30 American states had laws requiring the sterilization of those assumed to be genetically unfit, and following *Buck v. Bell,* many other states were encouraged to follow suit and to begin large-scale sterilizations. More than sixty thousand people were sterilized as a consequence of these laws.

By the mid-1940s, eugenics fell out of favor through much of the world. The Great Depression made people more certain that poverty was not simply the result of poor genes; the horrific eugenics policies of Nazi Germany, influenced strongly by the U.S. eugenics movement, started with sterilization and anti-immigration policies but ended with the mass murder of million of Jews and Romany (Gypsies). The Nazis' program took eugenics to its logical extreme and caused most of the world to condemn many of the assumptions upon which eugenics was based. But views that the genes determined individual and racial differences in behavior, personality, intelligence, and other complex traits did not disappear. Following World War II, the belief that genetics was an important consideration in future population improvement remained influential in small scientific circles. Additionally, as we will see later, the spectacular successes in genetics after the discovery of the structure of DNA encouraged an increased use of genetic principles in understanding many complex human characteristics—beliefs that continue to this day.

Figure 4.6: Carrie Buck with her mother, Emma. Arthur Estabrook Papers, M. E. Grenander Department of Special Collections and Archives, University at Albany, SUNY.

CONCLUSION

The growth of Mendelism triggered a major change in how heredity was understood and studied. While there remained a major divide between the Darwinians and the Mendelists for several decades following the rediscovery of Mendel's paper, by the 1930s the synthesis of Darwinism and Mendelism resulted in a new way of conceptualizing the relationship among heredity, development, and evolution. Mendelian genes were understood to be the units of heredity, and inheritance was seen as a law-like, predictable, and potentially powerful method for influencing the outcomes of reproduction, whether for plants, of interest to groups like the American Breeders Association, or for humans, of interest to the eugenicists.

Mendel's original research, his subsequent rediscovery, and the debates between the Darwinians and the Mendelians illustrate that the rise of a new conception of heredity was much more complicated, and much more interesting, than the simplified story suggests. This complexity is further illustrated by examining, in the next chapter, another major development that revolutionized our understanding of heredity: Thomas Hunt Morgan's work on the chromosomal theory of inheritance.

5

THE GENE AND UNIFYING BIOLOGY

INTRODUCTION

By the early twentieth century, several branches of biology had become quite prominent: Darwinism and its associated variety of explanations for evolution (Chapter 3); experimental embryology and cell biology, as exemplified by the work of Weismann and others who discovered the chromosomes and reduction division during gamete formation (Chapter 3); biometrics, the statistical study of populations (Chapter 4); and the new science of Mendelism, defined by William Bateson and his colleagues, which purported to explain the functioning of what came to be called the "gene" (Chapter 4).

We should be careful, however, in describing Darwinism, embryology, biometrics, and Mendelism as "several branches of biology." In fact, these activities were more or less independent of each other in terms of their interests and practitioners. Darwinism and Mendelism in particular were often seen as completely incompatible. Rather than understanding these as different branches of biology, it is better to think of them as separate early-twentieth-century research practices that focused on life's processes.

From our modern perspective, each of these is seen as a component of biology: genes, located on chromosomes, are the material basis of evolution; populations evolve as a result of natural selection acting on genetic variation, which itself is the result of the successive accumulation of random gene mutations (although, as described later, the modern view of the mechanism of evolution includes more than simply natural selection). But, in the early twentieth century, no such relationships were immediately obvious; in fact, while various practitioners of these fields had expressed the desire to unite them into one interconnected discipline—especially the Darwinists who, like Darwin himself, envisioned evolution and natural selection as holding the key to unifying the life sciences—the increasingly disparate conclusions of these practices discouraged most attempts to do so. Darwinism, in particular, was increasingly challenged as a consequence of the experimental findings of Mendelism and chromosome studies. The task of uniting these fields was accomplished only in the mid-twentieth century during what historians of biology have called the *modern synthesis*. This synthesis was a result of a combination of factors, including scientific discoveries, developments in mathematics, social developments, and philosophical commitments.

This chapter describes these developments, which spanned the first half of the twentieth century. As described, after Mendelian genes were located on chromosomes by Thomas Hunt Morgan and his colleagues and after a period of conflict between the increasingly influential Mendelism and the increasingly disparaged Darwinism, a growing commitment to make Darwinism more "scientific" and experimental combined with a conscious desire to unify the life sciences into a common theory of living systems, one unified science of *biology*.

THOMAS HUNT MORGAN AND THE PHYSICAL BASIS OF THE GENE

Thomas Hunt Morgan (Figure 5.1) and his team at Columbia University, in New York, discovered that genes were located on chromosomes. This discovery gave a physical reality to the hypothetical genes of Mendelism. Morgan thereby connected chromosomes to heredity, and cell theory to genetics.

In 1880–1886 Morgan studied zoology at the State College of Kentucky. From 1886 to 1890, he studied comparative morphology at Johns Hopkins University, in Baltimore, and from 1891 to 1904 he was a professor in experimental embryology at Bryn Mawr College, outside Philadelphia. In 1904, Morgan became a professor at Columbia University, where he discovered the hereditary function of the chromosomes. Morgan stayed at Columbia until 1928, at which point he founded the Division of Biological Sciences at the California Institute of Technology, where he remained until 1945.

In the late 1800s and early 1900s, Morgan, like most scientists, believed in evolution but was skeptical of natural selection. He also supported de Vries and Bateson's notions of discontinuous inheritance (see Chapter 4), but in fact he opposed

Figure 5.1: Thomas Hunt Morgan. Courtesy of the Archives, California Institute of Technology.

Mendelism itself, which he thought was too speculative and not in keeping with the growing popularity of experimental studies of nature, as exemplified by recent developments in experimental embryology. Morgan's own background in embryology led him to focus on physical material inside the cell as the basis of his research, and he was disinclined to place value on the sorts of abstract mathematical concepts with which the Mendelists dealt.

At Columbia, Morgan founded the *fly lab*, a research group that studied the fruit fly, *Drosophila melanogaster* (Figure 5.2). His students included Alfred Henry Sturtevant, Herman Joseph Muller, and Calvin Blackman Bridges. Sturtevant and Bridges entered the fly lab in 1910 as college juniors and became Morgan's graduate students in 1912. Both remained with Morgan at Columbia until 1928 and then relocated with him to Caltech. Muller completed his Ph.D. with the group from 1910 to 1916 and then left to pursue independent lines of research.

In 1908–1909, the group tried to find experimental evidence supporting de Vries' mutation theory of inheritance; it attempted to produce mutations in fruit flies by altering the flies' environments, in hopes of producing a new species as a result. While the group was unsuccessful in creating new fruit fly species, in the process it found evidence that sex inheritance was related to chromosomes; it described this evidence in a 1910 paper titled "Sex-Limited Inheritance in *Drosophila*." Morgan was not the first to notice a connection between sex-determination and the chromosomes; other researchers who studied cellular mechanisms in embryos had previously discovered that what we today call the X and Y chromosomes were associated with sex determination, such that XX embryos were female and XY embryos were male. But Morgan's discovery of this phenomenon in *Drosophila* spurred his interest in chromosomal behavior and led to further discoveries that located the source of heredity within the chromosomes themselves.

Morgan and his group subsequently discovered that eye color was linked to sex determination: they found male flies with white eyes rather than normal, or "wild-type," red eyes. They crossed this mutant fly with a wild-type (red-eyed) female fly, and the offspring were all wild-type. The group therefore assumed that the white-eye mutation was recessive, and when they interbred this group, they got the expected 3:1 ratios of wild-type and mutant offspring. However, they made a very unexpected observation: all of the mutant offspring were male. Since Morgan had already found evidence suggesting that the chromosomes were

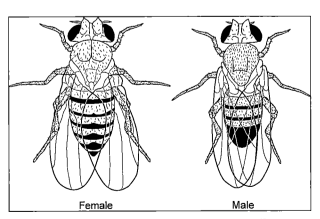

Female   Male

Figure 5.2: *Drosophila melanogaster,* the fruit fly. Illustration by Jeff Dixon.

Science and Technology: *Drosophila melanogaster*—A Model Organism

Drosophila melanogaster, the common fruit fly, has been an important contributor to the history of biology. The fruit fly is ubiquitous, found all over the world, and is especially common in major urban centers with large supplies of human agricultural products, especially fruits and fruit products, such as wine. Its ubiquity is a major reason for its popularity as an experimental organism.

But the early-twentieth-century introduction of the fruit fly to biology laboratories was not by design but by accident—or, more precisely, the fruit fly initially chose to inhabit laboratories and subsequently was put to work. Model organisms were (and are) a common feature of experimental biology, but the fruit fly was not considered especially useful as a model organism, unlike higher organisms that shared some physiological similarities to humans. Mice and guinea pigs were the preferred choice.

In 1901, William E. Castle used the fruit fly as a negative control for his model-organism breeding experiments. Castle was reluctant to use as a negative control animals that took time and energy to breed and grow; the fruit fly was thus a cheap alternative. Other experimental researchers turned to the fruit fly as a last resort, when other models simply did not seem to work well. Gradually, the fruit fly was seen as useful in its own right, particularly in studies of gene mutations. As easily observable mutations began to show up in these insects, their fame began to spread. Castle in particular introduced the fruit fly to an expanding network of biologists who wished to take a more experimental approach to the biological issues of their day, in particular breeding experiments and mutation studies.

Thomas Hunt Morgan and his fly lab made the fruit fly famous, and the group initiated a consistent and controlled study of the insect that itself became a model for later researchers. Following the successes of the fly lab, *Drosophila melanogaster* became a standard inhabitant of many genetics laboratories and remains so to this day.

involved in sex determination, this demonstration that eye color was also linked to sex convinced Morgan that the Mendelists' hereditary characters were located on chromosomes. This connection of Mendelism to physical material was what turned Morgan into a Mendelian, whereas before he had opposed it because of what he thought was its speculative nature.

For the next five years, Morgan and his team produced an enormous amount of experimental data describing the natures of and the relationship between genes and chromosomes. The "white-eyed" Mendelian character, or gene, was the first of many other characters that the group discovered by manipulating and breeding the fruit fly. A combination of a supportive academic environment, a productive experimental system (the rapidly breeding fruit fly), strong leadership in Morgan himself, and an able and cooperative group of researchers resulted in remarkable productivity within Morgan's fly lab.

By 1911, Morgan's group had produced evidence that many genes were located on the chromosomes, and at this point Morgan, still thinking in very material terms, suggested that genes were physically lined up on chromosomes in a linear fashion and that it might be possible to determine experimentally how far away these genes were from each other, or to *map* their distances apart. In 1909, Frans Alfons (F. A.) Janssens had discovered genetic crossovers, a process whereby two chromosomes (of a chromosome pair) actually exchange physical pieces of their structure during *meiosis,* the production of egg and sperm cells. The location of exchange of the chromosomal material can be anywhere, and the exchange occurs more or less randomly. Morgan suggested that genetic crossover could explain how two genes could be independently transmitted even if they were on the same chromosome: the genes would be separated if a crossover happened to occur between them. Morgan realized, in turn, that the likelihood of separating two genes depended on how close the two genes were to each other: if they were very close, the odds that a crossover would occur randomly between them would be correspondingly small. The further apart they were, the greater the odds of a crossover between them (Figure 5.3). Morgan realized that these odds could be quantified by studying the genes during fruit-fly breeding experiments.

Morgan looked to Alfred Sturtevant, one of his lab members, for a solution to the problem. Soon thereafter, in 1913, Sturtevant produced the first genetic *linkage map*, showing a spatial relationship between several genes on a chromosome, that was based on their frequency of co-inheritance (Figure 5.4). Within the next few years, he and others in the lab produced more genetic mapping data, illustrating the relative distances between many genes on chromosomes.

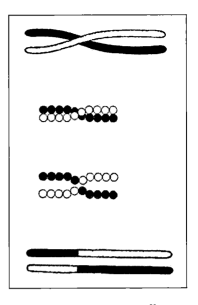

Figure 5.3: Morgan's 1916 illustration of a genetic crossover.

Morgan saw this genetic mapping work as lending material evidence for Mendelism. In 1915, he published *The Mechanism of Mendelian Heredity,* in which he described his group's mapping results and connected Mendelism to the behavior of chromosomes. Morgan also described results not just from the fruit fly but from other organisms as well, thus showing the broad applicability of Mendelian principles and of the physical location of genes on chromosomes. He gave Mendelism a physical rather than just a mathematical reality, and in the process he revolutionized heredity studies. Morgan advocated widely for Mendelism for the next decade, presenting his results in many different academic environments, publishing many articles describing additional experimental support, and producing several more books.

In 1928, Morgan moved his lab to the California Institute of Technology, an environment that was quite different from his fly lab. Caltech was part of the new

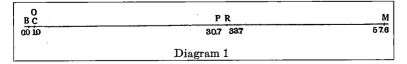

Figure 5.4: Alfred Sturtevant's illustration of the first genetic map.

wave of molecular biology that began sweeping through the biological sciences in the 1920s and 1930s and that we discuss in Chapter 6.

HEREDITY, DEVELOPMENT, EVOLUTION, AND THE GENE: THE MODERN SYNTHESIS

Introduction

Morgan's work stimulated a greatly increased interest in the gene, and the concept of the gene was increasingly seen as a potentially revolutionary means to understand some of the most fundamental questions in the major life science disciplines of the second quarter of the twentieth century. A diverse set of fields came under the influence of genetics in the first half of the twentieth century, including embryology, natural history, and, especially, Darwinism.

Embryonic Development: Hans Spemann and the Organizer

In embryology, genes were increasingly seen as the primary means to explain the complexity of early development in an organism. Much of this complexity was illustrated early in the twentieth century by the work of Hans Spemann, a German embryologist who revolutionized the study of development by the 1920s with his precise experimental work that demonstrated what he called *embryonic induction.* Over several decades, Spemann had completed a series of elegant experiments demonstrating that, within the developing embryo, there exist specific areas of embryonic tissue that could cause, or *induce,* other tissues to develop into specific organs. Spemann illustrated that, remarkably, these *inducer tissues,* when transplanted to new regions of the embryo, will cause organs to form in regions where they normally do not. Spemann also demonstrated that the *induced tissue* could itself become an inducer, triggering changes in yet another set of embryonic tissues; as a result, development appeared to be a cascading series of inductions. Spemann also suggested that there existed in the early embryo a primary inducer that initiated the entire developmental process; he called this the *organizer,* and by the late 1920s Spemann had completed several experiments suggesting that it was found in the *dorsal lip of the blastopore,* a region found in a very-early-stage embryo called the *blastula.* By transplanting this region to another young embryo, Spemann grew an entirely new organism in the region where the dorsal lip was transplanted (Figure 5.5).

By the 1930s, there arose many complicating factors in Spemann's conceptual understanding of embryonic development. For example, while it was initially assumed that induction was the result of chemical release by the inducer tissue, Spemann later found that induction could be initiated by dead tissues—and even contaminating dust seemed to be capable of induction in some of Spemann's experiments. Additionally, Spemann and others discovered that induced tissue could also have inductive effects on the inducer tissue itself. The process was thus seen as more complex, integrated, and nonsequential than was originally expected, and by the 1930s and 1940s very complex and sophisticated theories were developed that focused holistically on the developing embryo in an attempt to explain the experimental observations.

At the same time, a variety of researchers saw Morgan's work on the concept of the gene as potentially explaining the complexities of development. In 1934, Morgan himself wrote *Embryology and Genetics* as an exploration of these connections. While it was not until the 1950s and 1960s, with the revolutions in molecular biology and our understanding of the physical structure and function of the gene (see Chapter 7), that the genetic bases of development began to be understood, these earlier attempts strongly influenced a growing view that genetics and development were closely connected and that genes somehow directed development.

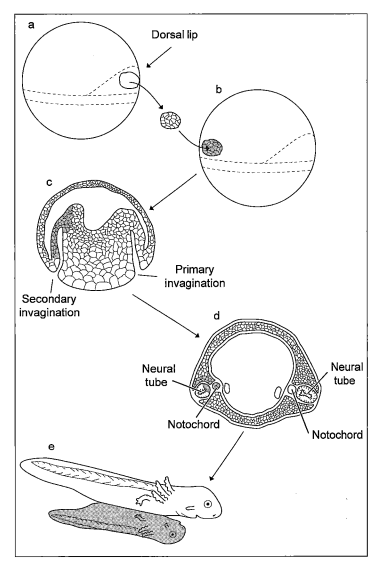

Figure 5.5: Hans Spemann's experiment demonstrating embryonic induction. Illustration by Jeff Dixon.

Population Genetics

The central focus of attempts to unify life science disciplines in the early twentieth century involved connecting Mendelism and Darwinism, the two critical components of the so-called modern synthesis. A key topic in the history of the modern synthesis is the origins of population genetics, a field of biology that applies mathematics to studies of the genetics of population change and evolution.

Mendel himself had of course used mathematics in his breeding experiments, discovering specific and consistent ratios of hybrid and of dominant and recessive pure-breeding offspring. Mendel's paper also offered a mathematical illustration of the process of *reversion;* over time, the frequency of hybrids in an inbreeding population decreased, and offspring reverted to the parental true-breeding forms. By the turn of the century, mathematically inclined Mendelians offered a corollary to Mendel's story: if population mating was random rather than inbred, Mendel's 1:2:1 ratio of offspring would be preserved, and these frequencies would be maintained in successive generations. G. H. Hardy and Wilhelm Weinberg, independently in England and Germany, respectively, demonstrated this mathematically in 1908. The work of the two has been summarized as the "Hardy-Weinberg equilibrium," a central concept in population genetics.

Hardy's and Weinberg's work was not especially influential in the first decade after its development. The rifts between Darwinism and Mendelism were sufficiently large as to discourage interest in relating Mendelism to large populations. It was difficult for those on either side of the debate to understand how Mendelism's precise ratios and the discrete, discontinuous changes in phenotype seen in his experiments (recall that Mendel's breeding of pea plants produced offspring with one of two forms of a given trait and no intermediate forms) had any relationship to population evolution and its apparent continuous evolution processes, with very minute, very gradual physical changes that required very large periods of time to produce observable effects.

But, by the 1920s, population genetics was coming into its own, led primarily by the work of Ronald Fisher, Sewall Wright, and John Burton Sanderson (J.B.S.) Haldane. These scientists were strongly interested in connecting Darwinism and Mendelism and saw the route to this synthesis as lying with mathematics.

Fisher was an early pioneer of many statistical tools still used today in population genetics and in statistics more broadly. He defined the statistical concept of *variance,* the square of the standard deviation, which describes quantitatively the dispersion of values away from the mean of a frequency distribution. He also introduced the concept of *genetic variance,* or *heritability,* a quantitative description of the amount of variation in a population that was due to genetic differences as opposed to environmental differences. He also introduced the notion that multiple genes, either independently, in an *additive* fashion, or by interacting with each other in an *epistatic* fashion, could together be responsible for producing one physical trait. These concepts of *epistasis* and *additive genetics,* defined by Fisher, remain central features of population genetics.

In a 1918 paper titled "The Correlation between Relatives on the Supposition of Mendelian Inheritance," Fisher first introduced his view of how Mendelism could solve a major problem of the Darwinian evolutionary theory of natural selection, in which continuous evolution prevented the possibility that new mutations could become established in a population. Fisher's central argument in this paper was that many traits that that vary in a continuous fashion, such as human height, were not the result of continuous inheritance, as believed by Darwinists. Instead, they were the result of many discontinuous Mendelian factors—*genes*—that, individually, are of minor significance but that add together to produce a given trait. In this way, the gene, according to Fisher, was the primary unit of natural selection, which itself was the primary mechanism of evolution. By positing the gene, which acts in a discontinuous fashion, as the primary unit of natural selection, Fisher thus provided a solution to the old problem of the "swamping out" of new traits in a blending inheritance model of Darwinism (see Chapter 4).

In a 1922 article titled "On the Dominance Ratio," Fisher expanded on what he perceived as the relationship between the discontinuities of Mendelism and the apparent continuities of phenotypic variation seen in large populations. Fisher argued that evolutionary change is a slow, regular process of natural selection acting in large populations on many additive genes, each of which contributes, independently and only in a small amount, to the heritable variation of any given trait. Fisher thus made *gradualism,* the doctrine of slow and extremely small changes espoused by Darwinists, compatible with the discontinuity observed in Mendelian breeding experiments. He also created a model for evolutionary change, with the units of selection being independent and equal, that was highly amenable to statistical analysis. He also argued that chance, such as random fluctuations in gene frequencies in small populations or drastic changes in gene frequencies caused by reproductive isolation of subpopulations, does *not* play a significant role in evolution.

Aside from this first description of how Mendelism could save Darwinism from the complications of continuous inheritance, Fisher contributed other central insights into the activity of genes from a population perspective. In 1928, Fisher published an article titled "The Possible Modification of the Response of the Wild Type to Recurrent Mutations," in which he outlined a theory of the evolution of the dominant-recessive relationship of alleles at a gene locus. Fisher hypothesized that if a new mutation arose in a population of organisms, it would have the best chance of remaining present in the population if it, in effect, hid itself by being masked by a normal allele. Selection would encourage any new, potentially deleterious mutation to be recessive; otherwise, it would quickly be selected against. A hybrid organism containing a normal and a mutant copy of the gene would have the best chance of survival if it phenotypically resembled the wild-type organism. Natural selection therefore selected those hybrids in which the effect of the new mutation was hidden, or recessive to the wild-type allele.

In 1930, Fisher published *The Genetical Theory of Natural Selection,* which summarized much of his theoretical work up to that point and which, as is evident from the title, explicitly advocated for the importance of the Mendelian genes as the fundamental unit of evolution by way of natural selection. In a later paper defending natural selection against its critics, Fisher stated that one of the main purposes of his book was "to demonstrate how little basis there was for the opinion . . . that the discovery of Mendel's laws of inheritance was unfavorable, or even fatal, to the theory of natural selection." *The Genetical Theory* described how Mendelism solved the swamping problem of Darwinist theories of continuous evolution, defended the importance of gradual evolution, and even extended much of his earlier work to human populations.

Unlike Fisher, Sewall Wright became interested in population genetics from a background in traditional biology. Wright was trained in physiological genetics under William Castle, who himself had carried out influential work on Mendelism and natural selection. Wright studied the genetics of coat color in guinea pigs and found that the trait appeared to be influenced by multiple genes interacting in an epistatic manner. He continued to work on coat color while at the Animal Husbandry Division of the U.S. Department of Agriculture (USDA, 1915–1925), where he developed his method of path coefficients, which allows analysis of causal relationships between correlated factors and the evaluation of the importance of one cause from multiple causes of a given effect. The method has become a central tool in many different areas of study involving quantitative analyses.

Also while at the USDA, Wright discovered the phenomenon of *fixation,* whereby highly inbred groups of guinea pigs exhibited a general decline in fitness, as characterized by factors such as low birth weight, high birth mortality, lack of vigor, and reduced weight. However, by crossing two highly inbred lines, Wright was able to improve their fitness dramatically. Wright explained these results by arguing that numerous homozygous recessive mutations cause a general lack of fitness in inbred lines. Consequently, according to Wright, "crossing results in improvement because each . . . [inbred line] supplies some dominant factors lacking in the [other line]." In keeping with his belief in the importance of epistasis, Wright hypothesized that the increased fitness seen in the hybrids resulted from the complex interaction of the increased number of dominant alleles.

These experiments were of fundamental influence on Wright's theories of evolution. Wright believed, for example, that smaller populations were the major focuses of natural selection. Such a view was in direct opposition to

Fisher's view of the importance of small, additive gene effects in large populations. Wright's interest in the role of smaller populations in evolutionary change is best exemplified by his view that *genetic drift,* the random fluctuations of gene frequencies that are characteristic in small populations with high rates of inbreeding, was a predominant factor in evolution. Wright argued that evolution would occur most rapidly in smaller populations because a new genetic mutation, if it arose in such a population, would have a better chance of randomly spreading through the population as a result of, for example, the random isolation and inbreeding of population members containing the new mutation. Such a mutation could therefore increase in frequency without requiring natural selection.

For Wright, therefore, not all evolutionary change was adaptive. Often, it was random. Wright disagreed with what he saw as Fisher's statistical simplification of the nature of animal populations; Wright argued that most animals live in smaller, scattered populations with significant reproductive isolation from other groups, and so chance plays a major role in changes in gene frequencies within such groups. In a 1930 review of Fisher's *The Genetical Theory of Natural Selection,* Wright outlined his own theory in contrast to that of Fisher. He reiterated his argument that small populations with relatively higher frequencies of inbreeding were centrally important to evolution. He also criticized Fisher's fundamental theorem of evolution for its neglect of dominance and epistatic effects and for its assumption that genes can be considered to contribute equally and additively to fitness. Wright also argued that genetic drift was crucial to evolution, by frequently creating new gene combinations on which natural selection could act:

> At a certain intermediate size of population ... there will be a continuous kaleidoscopic shifting of the prevailing gene combinations, not adaptive itself, but providing an opportunity for the occasional appearance of new adaptive combinations of types which would never be reached by a direct selection process. (Wright, 1930, p. 354)

This is the essence of Wright's major contribution to population genetics: his *shifting-balance* theory of evolution, in which random changes in gene combinations (i.e., genetic drift) can come to predominate and can become *fixed* in a population in the absence of natural selection. Wright did not rule out that such random combinations can also occasionally be adaptive and favored by natural selection, but natural selection was not essential to the process. Wright's theory is in direct contrast to Fisher's argument that the effects of genetic drift were negligible and that evolution was primarily the result of the small, equal, and additive effects of many adaptive genes. This conflict, between the relative importance of random events and of natural selection, and of large and small populations, has yet to be resolved in population genetics.

Like Wright, J.B.S. Haldane came to population genetics from an interest in Mendelism. Unlike Wright, however, he was never formally trained in biology. Haldane was educated at Eton (1908–1911) and at Oxford (1991–1914), in the United Kingdom, where he studied primarily mathematics, classics, and philosophy; with the exception of one lecture series at Oxford, he was never formally educated in biology. However, he developed a keen interest in Mendelian genetics; when he was eight, his father, John Scott Haldane (1860–1936), a physiologist at Oxford (1887–1936), took him to a lecture on Mendelism. As a teenager at Eton, Haldane and his sister Naomi (1897–1999) performed guinea pig, mouse, and rat breeding experiments. In 1915, in an article titled "Reduplication in Mice," Haldane, Naomi, and a fellow student, Alexander D. Sprunt, published one of the first cases of genetic linkage in mammals. In 1919, Haldane published two additional papers that attempted to explain discrepancies in gene linkage studies performed by Morgan and others.

Following his military service (1914–1919) during World War I, Haldane received a professorship in physiology at Oxford (1919–1921), following a brief tutorial on the subject from his father. He received a readership in biochemistry at Cambridge (1921–1933) and then moved to University College in London (1933–1957). At University College, Haldane was the Chair in Genetics and later in Biometry. In 1957, Haldane emigrated to India as a protest against the Anglo-French invasion of Suez; he worked for the Indian Statistical Office in Suez and then moved to Orissa, in India, where he set up a laboratory for genetics and biometry. He remained in Orissa until his death in 1964.

Between 1924 and 1932, Haldane published nine papers under the collective title "A Mathematical Theory of Natural and Artificial Selection." These were intended to quantify the natural selection process. Haldane began by testing mathematically the results of selection on the gene frequencies of successive generations of simple Mendelian populations. Each paper explored this question under a different set of parameters.

From these papers, Haldane developed his theory of evolution. Like Fisher, Haldane argued that selective processes acted upon individual genes in large populations, in an additive process. Unlike Fisher, however, he argued that single mutations and chromosomal variation could have a very large selective influence. Haldane also advocated a theory of dominance different from that of Fisher or Wright: he argued that dominance resulted from the selective increase of wild types to compensate for the appearance of inactive mutants. Haldane also developed a theory of the selection of favorable gene combinations that was similar to Wright's shifting-balance theory but that placed more emphasis on the role of natural selection. In 1932, Haldane published *The Causes of Evolution,* which summarized his theory of evolution by natural selection and compared it to those of Fisher and Wright. Like Fisher's *The Genetical Theory of Natural Selection,* the book was intended to show that Mendelism did not discredit, and indeed was complementary to, Darwinism.

The work of Fisher, Wright, and Haldane was pivotal in resolving one of the most contentious issues of their generation: the conflict between the Mendelists, who supported discontinuous evolution, and the Darwinists, who supported continuous evolution. In arguing that Darwin's apparently gradual process of natural selection could be explained with reference to discrete Mendelian characters, or *genes,* Fisher, Wright, and Haldane illustrated the compatibilities of Mendelism and Darwinism. This compatibility was further addressed by a new generation of biologists in the 1940s who explicitly worked to unify biology, defining what has come to be known as the *modern synthesis.*

Neo-Darwinism and the Modern Synthesis

By the 1930s, Darwinism had come under increasing attack. The rise of Mendelism had resulted in strong support for the notion of discontinuous inheritance, or abrupt changes in physical form from one generation to the next. While those who studied the natural history of populations pointed out that most populations actually exhibit a continuity of variation when observed in their natural habitat (for example, organism height exhibits a great deal of continuous variation) and that Mendelism failed to account for this, the popularity of the experimental approach to nature was such that these criticisms were not sufficient to defend continuous-inheritance theories against Mendelism.

Following the work of Fisher, Wright, and Haldane, however, there grew an increasing interest in reviving Darwinism from its languishing in the early twentieth century. Several influential individuals wrote very explicitly about a perceived need to revive Darwinian evolution using Mendelian principles for the purpose of using Darwinism as the central feature of a new, unified concept of biology.

One influential figure was Theodosius Dobzhansky, a geneticist trained in Morgan's laboratory who, influenced by Wright's shifting-balance theory, began studying the population genetics of wild populations of the fruit fly. Dobzhansky confirmed that genetic drift was a major factor in fruit-fly evolution, but he also found that natural selection played a major role. However, and like many of his contemporaries, Dobzhansky gradually came to view natural selection as the more important agent of evolutionary change. Dobzhansky's very influential book, *Genetics and the Origin of Species* (1937), defined evolution, and the closely related subject of the formation of new species, in terms of changes in gene frequency in populations. Following the lead of the earlier population genetics, he described evolution at the level of the population and emphasized the importance of mathematics to studying the topic. Dobzhansky helped translate the often difficult language of population genetics into concepts more easily understood by experimental geneticists and thus further encouraged the connections made between Mendelism and Darwinism.

Dobzhansky was the first of a number of individuals who, at approximately the same time period, published very influential books that applied Mendelism to evolution and, in the process, rescued Darwinism from obscurity. Cyril Darlington's *Recent Advances in Cytology* (1931) and *The Evolution of Genetic Systems* (1939) and Michael White's *Animal Cytology and Evolution* (1945) described how genes were involved in the production of variation upon which natural selection could act. Julian Huxley's *Evolution: The Modern Synthesis* (1942), in which the author coined the phrase *modern synthesis*, and Ernst Mayr's *Systematics and the Origin of Species* (1942) argued that the formation of new species could be explained by successive small changes in gene frequencies over sufficiently vast periods of time. George Gaylord Simpson, in his book *Tempo and Mode in Evolution* (1944), argued, in contrast to many paleontologists of his day who were still strongly influenced by Lamarckian notions of progressive evolution (side definition: a paleontologist studies the fossil record in order to understand evolution), that natural selection was generally sufficient to explain the evolutionary changes that could be observed in the fossil record (although, in some cases where large gaps existed in the fossil record, Simpson appealed to Wright's conception of genetic drift).

CONCLUSION

By the late 1940s, the modern synthesis had successfully revived Darwinism by explaining it in genetic terms and by then applying this new "neo-Darwinian" understanding to such studies of life as population biology, paleontology, classification, and species formation. Evolution was explicitly seen as a way to unite all of biology under one central theme. As Dobzhansky argued in a 1973 article, "Nothing in biology makes sense except in the light of evolution." And evolution itself did not make sense to these biologists except "in the light" of the concept of the gene.

Genetics rescued Darwinism from the perils of continuous inheritance, even though originally it was responsible for actually causing its demise. By using genetics as an explanation for the mechanism of evolution, where evolution was defined in terms of the dynamic change of gene frequencies in populations, Darwinism and natural selection became, in the mid-twentieth century, a central explanatory strategy in the science of biology.

But this dominance of Darwinism came under increasing criticism as an overarching explanatory framework for all of biology. In particular, there remained a number of questions that natural selection could not answer. Embryology remained outside the new unification, despite some gradual influence of genetics in explaining development. Additionally, microbial evolution and the origins of multicellular organisms were still not understood in any meaningful detail. But these topics became understood as biologists came to understand the nature of the gene. The modern synthesis was to be fundamentally transformed by the rise of molecular biology, a field of practice that, in the wake of Morgan's illustration of the physical nature of genes on chromosomes, focused on discovering exactly what genes were and how they worked.

6

MOLECULAR BIOLOGY AND THE GENE

INTRODUCTION: THE SEARCH FOR THE GENE

In the previous chapter we described the modern synthesis and the work of Thomas Hunt Morgan, who demonstrated that the "hereditary characters" of Mendelism appeared to have an actual, physical, material basis on the chromosomes. We also noted that Morgan developed linkage analysis, a technique that could be used create genetic maps showing the relative locations of genes on a given chromosome. But Morgan and his group did not work with actual, physical genes. They studied genes indirectly, finding their approximate locations on chromosomes using linkage analysis. Although he had assumed that genes were a physical entity, Morgan did not actually locate and study the genes themselves.

But Morgan's demonstration that genes were located on chromosomes stimulated many attempts to understand the physical reality of genes and the mechanisms by which they functioned—that is, what they were and how they worked. By the turn of the century, chromosomes were known to be a mix of mostly protein and another molecule called deoxyribonucleic acid (DNA), first discovered in 1868 by the Swiss biologist Friedrich Miescher and subsequently ignored (more or less) until the late 1940s. Following Morgan's work, most of the research focus on what constituted the gene came to rest on proteins. From the 1930s to the early 1950s, this work, dubbed *molecular biology,* dominated the new experimental biology.

THE BEGINNINGS OF MOLECULAR BIOLOGY

The term *molecular biology* was coined by Warren Weaver, Director of the Division of Natural Sciences at the Rockefeller Foundation from 1932 to 1955, in the Foundation's 1938 Annual Report. Weaver defined molecular biology as the study of molecular or subcellular structures. This early form of molecular biology is not identical to that which is described as molecular biology today. Modern molecular biology refers to research involving the analysis and manipulation of DNA as the physical basis of inheritance, whereas most molecular biological research of the 1930s and 1940s focused on proteins. However, this earlier biochemical tradition of molecular biology is a direct ancestor of today's field. Weaver's molecular biology program created the institutional, organizational, and intellectual foundations of modern, DNA-based molecular biology. Additionally, Weaver's interests, reflected in the Rockefeller Foundation's philanthropic efforts under the "Science of Man" project to improve society by improving human biology, centered around attempts to understand the physical basis of heredity—that is, in the wake of Morgan's research that localized the genes to the chromosome material, it focused on and stimulated research that attempted to discover what, physically, constituted the gene. As mentioned earlier, the primary molecule of interest became the protein, but the goal, the search for the gene, seen as the "secret of life," was the same.

The Rockefeller Foundation and the Birth of Molecular Biology

It is commonly accepted that the Rockefeller Foundation in the United States played a pivotal role in the creation of molecular biology as a discipline from about 1920 to 1950, funding biological research in the hope that it would be applicable to addressing many issues of social importance and concern. John Rockefeller founded the Rockefeller Foundation, in 1913, as a philanthropic organization, with the goal of funding various sciences believed to be of important benefit to humanity. Eugenics was one of the most important of these sciences, and the Foundation heavily funded eugenics work. By about the 1930s, explicit research programs in eugenics were less popular, but the Rockefeller Foundation continued to be interested in harnessing the social and biological sciences to control perceived social ills. The latest cutting-edge techniques in what came to be called molecular biology promised a much more efficient, sounder, although considerably slower, method of directing human society's future development, than traditional eugenics.

Warren Weaver became Director of the Natural Sciences Division of the Rockefeller Foundation in 1931, to implement what was called the *Science of Man* project: a focus on funding research that could be applied to human social issues. The project, with its explicitly stated goal of "social control," was unveiled at a 1933 planning meeting of the Rockefeller Foundation's trustees and officers:

Science has made significant progress in the analysis and control of inanimate forces, but science had not made equal advances in the more delicate, more difficult and more important problem of the analysis and control of animate forces. This indicates the desirability of greatly increasing emphasis on biology and psychology, and upon those special developments in mathematics, physics and chemistry which are themselves fundamental to biology and psychology. . . . The challenge of this situation is obvious. Can man gain an intelligent control of his own power? Can we develop so sound and extensive a genetics that we can hope to breed, in the future, superior men? Can we obtain enough knowledge of the physiology and psychobiology of sex so that man can bring this pervasive, highly important, and dangerous aspect of life under rational control? Can we unravel the tangled problem of the endocrine glands, and develop, before it is too late, a therapy for the whole hideous range of mental and physical disorders which result from glandular disturbances? Can we solve the mysteries of the various vitamins so that we can nurture a race sufficiently healthy and resistant? Can we release psychology from its present confusion and ineffectiveness and shape it into a tool which every man can use every day? Can man acquire enough knowledge of his own vital processes so that we can hope to rationalize human behavior? Can we, in short, create a new science of man? (Quoted in Kay, 1993, p. 45)

The Science of Man program rested on the belief that progress in science and civilization had outpaced humanity's ability to deal with scientific findings wisely and that a "better man" was needed to ensure the future health of the human race. By this time the Rockefeller Foundation had rejected a "crude" eugenics approach to the goal of improving humanity, but the goal itself had not changed: the Science of Man program was created in order to find a new way to scientifically approach social ills, using a more sophisticated eugenics. Weaver and others envisioned that researchers from various disciplines, especially physics, chemistry, and math but not so much traditional fields of biology such as evolution, would become involved in this new way of studying vital phenomena, for the explicit and directed purpose of achieving goals laid out in the Science of Man project. At its inception, therefore, molecular biology had as its ultimate goal a technologically mediated improvement of humanity, with the assumption that such direct biological intervention would result in social progress.

WHY PROTEINS?

One important reason for the focus on proteins in the early search for the physical basis of the gene was the growth of biochemistry in the early twentieth century. Biochemistry refers to the study of chemical reactions inside the cells of organisms. Biochemistry, the application of chemistry to biology, arose in the context of the rise of chemistry as a central and politically relevant science. Chemistry became a prominent discipline during World War I, when it was used for a variety of war-related purposes, including elucidating the nature and importance of vitamins and providing nutritional advice to troops, producing raw materials for creating war materials such as ammunition, and producing chemical weapons. Following World War I, biochemistry became increasingly influential in the field of medicine, supplanting the traditional dominance of physiological chemistry in medicine (chemical reactions in organisms but not in cells, including processes such as food digestion and body excretion and the composition of the blood). Biochemistry also benefited after World War I from continued industrial and state interests in harnessing, and eventually imitating, the chemical properties of living things (especially plants). By the 1920s and 1930s, biochemistry was an institutionalized interdisciplinary practice of major social importance.

Central to the rise of biochemistry was the discovery of the existence of organic polymers. Jons Jacob Berzelius, in addition to his many other achievements in chemistry, coined the term *polymer* in 1833 to refer to compounds that had similar proportional chemical structures but different molecular weights. For example, the compounds CH_2, C_2H_4, C_3H_6, C_4H_8, and so on all have the same proportion of elements (one carbon for every two hydrogens) but different weights. Polymer chemistry became an important field of study in the early twentieth century as polymers were increasingly exploited for industrial uses.

An equally important concept in the early twentieth century was the *colloid,* a name used to describe polymers as being constructed of aggregates of small molecules that fit together to form larger structures; it was assumed that polymers were themselves simply too large to be proper, stable molecules but instead were smaller molecules held together by an unknown substance believed to be unique to living things. Colloidal theory dominated in the first two decades of the twentieth century. Although we now know that colloid theory is incorrect, since polymers are indeed large molecules, it was a very successful theory in its day, leading to large bodies of research and attempts to establish departments and even schools of colloid science.

But by the 1930s colloid theory had been replaced by the theory of *macromolecules,* which argued that polymers were in fact very large molecules. While at first glance macromolecular theory seems to rid biochemistry of the vitalist properties of colloids (which, as just mentioned, were assumed to be held together by an unknown substance unique to living things), in fact both theories fulfilled a similar role of keeping a role for vitalism in contemporary understandings of life. Enzymes are, after all, also active and unique to living things. Both macromolecules and colloids were thus important concepts introduced in order to address a central issue at the time: how could one reconcile the uniqueness of life with the growing understanding of physics and chemistry? Both colloids and macromolecular proteins such as enzymes were simultaneously chemical and physical substances *and* unique to organisms. This struck a balance between the competing mechanist and vitalist philosophies as explanations for life.

Enzymes were a central molecule of interest under macromolecular theory; these were conceptualized in the 1900s as special proteins that could activate metabolic processes in organisms. By the 1930s, many laboratory techniques were created for studying enzymes *in vitro* (outside the body). Additionally, central biochemical reactions, all driven by the activity of enzymes, were discovered during this time, including glycolysis and the Krebs Cycle (in which sugar is broken down to be used for energy), the urea cycle (in which by-products of amino acid metabolism are eliminated), and photosynthesis (the conversion in plants of carbon dioxide and water into sugar and oxygen, using energy from the sun), among others. By the 1930s, biochemistry was a well-established discipline with many central achievements that have become the foundations of modern biochemistry.

The central object of study of biochemistry was the enzyme, and when early molecular biologists began to wonder about the physical structure of the gene, the findings of biochemistry lent credence to the widely held assumption that the gene was some sort of enzymatic protein. Chromosomes are themselves a mix of mostly protein with some nucleic acid (DNA). Protein was the obvious candidate to most biologists in the 1930s and 1940s; by the 1930s, DNA was known to consist of subunits of sugar phosphates and four other molecules called bases (guanine, adenine, thymine, cytosine), and, while its macromolecular structure was unknown, it was widely assumed to be a uniform and monotonous repetition of subunits, each containing the four nucleotides present in DNA. With such a configuration, it seemed difficult to conceive of DNA as encoding the totality of all proteins present in an organism. DNA was assumed to be at most a minor helper molecule of some sort. In contrast, proteins were known to include the very reactive enzymes, capable of stimulating and sustaining very specific cellular activity. They were also known to be composed of about 20 amino acid subunits, and many different types of proteins had been discovered that were capable of carrying out a large variety of functions. From the 1930s to the 1950s, therefore, protein research dominated molecular biology.

FEATURES OF EARLY MOLECULAR BIOLOGY

Like other emerging disciplines in the early twentieth century, molecular biology emphasized the unity and commonality of life rather than its diversity; therefore, phenomena common to all life, such as heredity and cellular processes, were studied in the simplest living organisms and findings extrapolated to more complex organisms such as humans. The use of animals as so-called models of human functioning was also commonplace. Molecular biology was also reductionist; that is, it involved the search for general physical and chemical laws underlying life, ignoring as a research focus higher explanatory frameworks such as comparative anatomy and physiology or interactions with the environment and with other organisms. Molecular biology instead focused on how the simplest physical and chemical laws governed life's complex processes.

Other emerging life-sciences disciplines also focused on the chemical processes of life, and in this sense molecular biology was not unique. As discussed earlier, biochemistry, for example, focused on the basic chemical reactions that occur inside the cells of organisms. And both biochemistry and molecular biology focused on proteins as their molecule of interest. But, although both biochemistry and molecular biology were reductive and had proteins as their research focus, the two fields were otherwise very different. Biochemistry focused on the *activity* of proteins (and enzymes in particular); its interest was in chemical reactions. In contrast, molecular biology's interest in proteins lay in its search for the gene. Proteins, for the molecular biologists, were the secret of life, and this secret was somehow locked in their *structure*. Molecular biology was thus a science of structure rather than activity, with molecular biologists searching for clues about how molecular structure carried hereditary information.

Molecular biology also differed from biochemistry in that it ignored disciplinary boundaries, using methods (and encouraging collaborations and disciplinary career changes) from many branches of biology and from physics, chemistry, and mathematics. It also used complex techniques and technologies for the study of macromolecular structure, such as electron microscopes, ultracentrifuges, electrophoresis, spectroscopy, x-ray diffraction, and radioactive isotopes. Molecular biology researchers typically worked in large research teams managed by strong scientific leaders, sharing expensive equipment and collaborating with various specialists on the team.

MOLECULAR BIOLOGY INSTITUTIONS: CALTECH

The Rockefeller Foundation provided enormous grants to several American research institutions for the purpose of building strong molecular biology research centers. One of the most important of these was the California Institute of Technology (Caltech), which is an excellent example of how the new molecular biology worked and how the Rockefeller Foundation guided its formation.

Caltech was the best-funded of all the institutions funded by the Rockefeller Foundation, as it was thought to be an ideal environment for seeding a molecular biology research program. It had no previously established biological traditions such as medical education, agricultural interests, evolutionary biology, or natural history to compete with the emerging molecular biology program; it was interdisciplinary, focusing on a problem-oriented approach using multiple approaches to problem-solving; many of Caltech's leaders had close ties with industry and philanthropic organizations (Warren Weaver at the Rockefeller Center was himself a faculty member at Caltech); and it was located in Southern California, a major industrial center focused on developing large, highly coordinated managerial structures. Researchers from all over the world came to visit Caltech to learn the new techniques and research strategies of molecular biology. Hundreds of graduate students were trained there, propagating the new methods to other research environments. Caltech helped seed the molecular biological approach, training a first generation of researchers and providing a model for the growing discipline.

Some of the leading scientists practicing the new molecular biology were located at Caltech. These molecular biologists often became leaders of large research groups. Using money from the Rockefeller Foundation and similar organizations, they built large, diverse teams of researchers that focused on common problems, using people and methods from multiple disciplines. Thomas Hunt Morgan founded Caltech's Division of Biology in 1928 and remained its head until his death, in 1945. Upon his arrival at Caltech, he set about building a research program consisting of Departments of Genetics, Embryology, Physiology, Biochemistry, and Biophysics. By the mid-1930s, Morgan was famous for both his research and his leadership qualities, having won a Nobel Prize in 1933 for his chromosome work and being widely admired for his leadership of the legendary fly lab. His Caltech institution-building was of prime importance in establishing the molecular biology style advocated by the Rockefeller Foundation.

Linus Pauling, an American chemist, was another strong and influential Caltech leader. Pauling became a member of Caltech's chemistry division in 1931 and was its Director from 1937 to 1964. By the 1930s, Pauling was the foremost researcher studying the nature of the chemical bond, and by the mid-1930s he began focusing heavily on applying his work to elucidating the structure of proteins, largely as a direct consequence of the Rockefeller Foundation's influence and funding. Pauling summarized much of his work in a 1939 publication titled *The Nature of the Chemical Bond,* one of the most celebrated books in the history of modern chemistry. Pauling was a bold, aggressive, ambitious scientific manager who revolutionized protein research and who won a Nobel Prize for this work in 1954.

Max Delbrück, a German physicist who previously had studied atomic physics, worked in Morgan's Caltech biology division as a Rockefeller fellow from 1937 to 1940. Upon his arrival at Caltech, Delbrück became interested in viruses, or *phage*, which he conceived of as the atoms of biology: the simplest units of life. While at Caltech, Delbrück began a phage research program. In 1940, he assumed an Associate Professorship at Vanderbilt University. While at Vanderbilt, in the early and mid-1940s, Delbrück established himself as a leader in research on phage, which were becoming a widely recognized tool for studying some of the most fundamental principles of molecular biology.

PHYSICISTS INVADE BIOLOGY: THE PHAGE GROUP

In 1941, Max Delbrück met Salvador Luria at the Cold Spring Harbor Symposium on Quantitative Biology, an annual summer research symposium begun in 1933 to discuss developments in molecular biology. Delbrück and Luria began what became commonly known as the *phage group,* an informal but very influential research group, spread across multiple institutions, that used phage to study gene function. By 1945 Delbrück began to lead a *phage school* at Cold Spring Harbor, which served to rapidly expand the number of researchers interested in phage. The phage group included among its members some of the most important figures in the history of molecular biology, notably James Dewey Watson, who, prior to his co-discovery of DNA structure (see Chapter 7), was a member while he was a promising young graduate student under the supervision of Luria at Indiana University from 1947 until 1950. The phage group exemplified many of the principles of molecular biology, working as large research groups, focusing on a very simple model organism, and explaining life processes using a reductionist approach. In 1947, Delbrück, at the height of his fame as the father of phage genetics, moved to Caltech to expand its phage research program.

Delbrück's connections to physics inspired many physicists to enter molecular biology shortly after World War II. Many of these physicists thought that most of the great discoveries in physics had already been made and that biology was the next frontier. Some were greatly disillusioned by the use of physics to produce nuclear weapons during World War II. Many saw the discovery of the chemical nature of the gene as the next great challenge in science. Thus, like biochemistry's growth during and after World War I, when the social prominence of chemistry influenced biology and produced biochemistry, molecular biology benefited significantly from a wartime context in which the prominence of physics led to its invasion of biology and the formation of molecular biology. The involvement of prestigious physicists helped to enhance the image of molecular biology as a cutting-edge academic discipline.

MOLECULAR BIOLOGY AND MACROMOLECULAR STRUCTURE

Physicists came to biology with a tendency toward reductionism, reducing the complexities of life to underlying chemical and physical processes, as had become increasingly common in the life sciences by the 1930s. But, in the case of molecular biology and its focus on structure, physicists were especially influential. For example, in 1944, Erwin Schrödinger, one of the chief architects of quantum mechanics, wrote *What Is Life?*, a short book that influenced many physicists to pursue molecular biology. Inspired primarily by Morgan's linkage studies and Delbrück's work with phage, Schrödinger linked physics and chemistry to biology by discussing heredity's requirement for molecular stability through multiple generations. *What Is Life?* is one of the first books to compare the gene to a code, somehow encoding information about heredity within the sequence of its subunits rather than its secondary, three-dimensional folding structure. But his primary focus was the gene's remarkable stability. Schrödinger argued that the gene was a unique marvel of reality, capable of maintaining orderliness in the inherently disordered universe of atomic-scale phenomena. Only the gene, Schrödinger argued, could avoid the natural state of random motion that is typical of natural phenomena at the atomic level: its combination of extremely small size and almost permanent stability was unique to life. For Schrödinger, the key to explaining life was explaining this orderliness and explaining how the hereditary molecule could remain stable over vast periods of generational time. Originally written for a popular audience in order to explain development and heredity in terms of the physical and chemical behavior of molecules, the book became a classic text influential in accelerating physicists' interests in the gene. Its publication marked a pivotal moment in the rapid growth of post–World War II molecular biology.

INSTITUTIONALIZING MOLECULAR BIOLOGY

While initially nurtured by the guidance and financial support of the Rockefeller Foundation and similar institutions in the United States and elsewhere, by the 1940s molecular biology became a science of major political interest. Especially following World War II, with the bolstering of molecular biology by the invasion of physics, molecular biology became increasingly institutionalized and recognized as a powerful new science, both in terms of its research potential and its connections to political and economic power. During World War II, government support for molecular biology grew dramatically, particularly in the United States but also elsewhere. In the United States in particular, molecular biology and its biomedical applications were increasingly funded through various governmental and nongovernmental agencies that were either created or expanded considerably during and immediately following the war. One of the most important governmental organizations was the National Institutes of Health (NIH), which was founded in 1887 but became the central biomedical sciences funding agency in the post–World War II period. In 1946, the NIH became a grants-administering agency, with a granting budget beginning at about $4 million but growing to more than $100 million by 1957 (and to $1 billion by 1974). The entire NIH budget expanded from $8 million to almost $1 billion in the same ten-year time period, 1947–1957. The organization grew from the singular National Institute of Health to the plural National Institutes of Health, growing to encompass 10 separate institutes by 1960, 15 by 1970, and 25 by 1990. The NIH remains today the central biomedical funding agency in the United States.

CONCLUSION

Beginning in the 1930s, following Thomas Hunt Morgan's influential and persuasive work that demonstrated the physical localization of genes on chromosomes, a new research program sought to identify the physical nature of the gene and to understand how it worked. In a climate of an increasing focus on reducing the complexities of life to physical and chemical processes, this new discipline, molecular biology, rapidly became a dominant scientific discipline in many countries, most notably the United States—which, following World War I, itself had become a dominant nation in the sciences. From its beginnings, molecular biology focused on the physical structure of molecules, in the hope that studies of structure would reveal the secret of life—that is, the means by which genes, as physical molecules, functioned chemically to determine biological development. This secret was assumed to reside in the structure of proteins, the dominant component of chromosomes. As a result of the rise of biochemistry and the study of enzymes, proteins were a popular molecule that seemed to be capable of innumerable metabolic functions, and it seemed logical that heredity might potentially be another of protein's activities.

Molecular biology resonated initially with the goals of major philanthropic organizations such as the Rockefeller Foundation in the United States, which, in the context of eugenic goals of biologically improving humanity, saw in the new science an exciting new means of social reform. Beginning with the support of the Rockefeller Foundation, molecular biology grew in the post–World War II period into a mature and powerful research discipline that dominated the biological sciences.

By the early 1950s, proteins were still the dominant focus of molecular biology research, but at about this time a new view began to emerge. A number of scientists became interested in that other component of chromosomes, what had for more than twenty years been assumed to be a lowly, inactive, helper molecule of simple structure and modest function: deoxyribonucleic acid, or DNA, the subject of the next chapter.

7

THE SWITCH TO DNA

INTRODUCTION

As discussed previously, it was widely assumed in the 1930s and 1940s that protein was the physical basis of the gene. By the mid-1940s, however, some molecular biologists began to focus on DNA, the other component of chromosomes. Several experiments performed at this time suggested that DNA itself was the molecular substance of the gene. But it was the creation, by a 35-year-old graduate student and a young, brash, and impatient postdoctoral student, of a molecular model made out of steel and plastic that revealed the structure of this molecule and suggested how it might function as a gene. Beyond this construction of a molecular model of DNA, the final determination of how DNA actually worked to produce proteins was the subject of an additional 15 years, approximately, of theoretical and experimental work by molecular biologists—but the solution was finally found by a biochemist.

EARLY EVIDENCE FOR DNA

In 1944, Oswald Avery, a microbiologist at the Rockefeller Institute for Medical Research in New York City, conducted an experiment that suggested that DNA, not protein, appeared to be the hereditary material. Avery used the isolated DNA from a bacterial strain to transform another strain—that is, to transmit some of its biological properties to another bacteria. Avery himself drew modest conclusions from the work, and others followed suit; his experiment was seen primarily as an interesting result warranting further research on DNA, and only a minority of molecular biologists became interested in DNA. Interdisciplinary differences also played a role in causing a delay in the acceptance of Avery's work: Avery was a microbiologist, and therefore an outsider in the domain of molecular biology.

James Watson, a student with Herman Muller at Caltech (as described in Chapter 5, Muller was originally part of Thomas Hunt Morgan's early fly lab at Columbia prior to being a professor at Caltech) and later a graduate student with Salvador Luria, was inspired by Avery's results to study DNA. Watson traveled to Copenhagen to study nucleic acid biochemistry. In 1951 he saw a presentation by Maurice Wilkins, a former physicist and research associate at King's College London, who used X-ray crystallography to study DNA structure. X-ray crystallography involves bombarding a structure with X-rays, which are bent or spread in different directions depending on the shape of the structure. Watson was inspired by the presentation, and he persuaded Salvador Luria to get him a postdoctoral position at the Cavendish laboratory in Cambridge, England, where X-ray crystallographic techniques were routinely used.

WATSON AND CRICK'S DNA MODEL

Watson began his postdoctoral work at the Cavendish Laboratory in the fall of 1951. Very shortly after his arrival at the Cavendish, Watson met Francis Crick, an ex-physicist who had been inspired by Schrödinger's *What Is Life?* to pursue a Ph.D. in biology. Crick was working on his Ph.D. when Watson arrived, analyzing X-ray diffraction patterns of DNA. Crick, like many others, was far from convinced that DNA was the hereditary molecule, but he did assume that it was an important component of chromosomes. But Watson's enthusiasm for DNA infected Crick, and together they began attempts to build a molecular model of DNA's structure.

By this time and shortly after, several other lines of research inspired and aided Watson and Crick with their model-building. In 1948, Edwin Chargaff discovered that DNA had an equal number of adenine and thymine bases and an equal number of cytosine and guanine bases. In 1949, Sven Furberg determined the three-dimensional structure of a single DNA nucleotide—the smallest structural unit of DNA. In 1951, Linus Pauling described the structure of a protein molecule using molecular models, inspiring Watson and Crick to try the same for DNA. In 1952, Alexander Todd discovered that the phosphates molecules of DNA are linked together in DNA into a chain.

In the same year, John Griffith discovered that adenine attracts thymine and guanine attracts cytosine, and Alfred Hershey and Martha Chase, two phage group members, showed that when a virus infected a bacterial cell, only its nucleic acid entered the cell, and this was sufficient for the production of new viral particles. The results of this experiment were in fact not as conclusive quantitatively as the earlier Avery experiment, but it was nevertheless more influential and more persuasive; the influence of the phage group helped make their conclusions more widely accepted.

The most important experimental work that influenced Watson and Crick was done by Rosalind Franklin, a young and recent new research associate at John Randall's laboratory at King's College London. Franklin was an X-ray crystallographer but new to DNA research. She arrived at King's in January 1951, assuming that she would have her own research program to analyze DNA structure. This was apparently Randall's intention, but Maurice Wilkins, in the same laboratory but away when Franklin was hired, thought of her simply as his assistant. Wilkins' view of Franklin's position contrasted with her own expectation for her role; the misunderstanding, eventually acknowledged but never really overcome, combined with dramatic differences in personality, resulted in tension between Franklin and Wilkins that was never resolved.

In November 1951, Franklin presented x-ray crystallographic data of DNA at a colloquium that Watson attended. On the basis of her data, Franklin suspected that DNA was a spiral chain of multiple strands with a sugar-phosphate backbone on the outside of the molecule, but at the time she felt her data did not yet prove this conclusively, and she did not yet know how the base pairs of DNA were bound to each other. Watson tried with Crick to construct a model based on Franklin's results, but by his own account he forgot most of the details and did not understand the rest because he had little knowledge of X-ray crystallography. As a result, Watson and Crick built a three-strand model with phosphate groups inside as a backbone. They showed their model to Franklin and Wilkins, who pointed out the obvious errors and dismissed the value of model-building. Watson and Crick were told to stop these efforts and to go back to their respective research projects.

In January 1953, Linus Pauling unexpectedly published a DNA model. This shocked Watson and Crick, who thought they were scooped, but in fact Pauling made what some have argued was a fairly obvious error: the phosphate groups were given the wrong electric charge. As a result, Pauling developed a three-chain model with a phosphate backbone, similar to Watson and Crick's prior model. Watson and Crick were relieved but felt a renewed pressure in the race for the structure. Watson spoke with Wilkins at Kings, who possessed the modeling equipment, to try to persuade him of the importance of model building. During this meeting Wilkins showed Watson Franklin's most recent X-ray, photo 51 (Figure 7.1), produced in mid-1952, that strongly suggested a two-chain model.

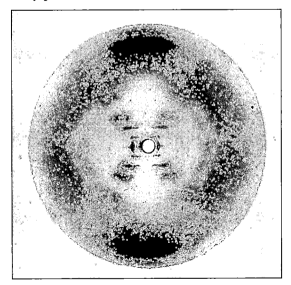

Figure 7.1: Rosalind Franklin's "Photo 51." Courtesy of Oregon State University Libraries, Special Collections.

Wilkins shared the data without Franklin's or Randall's consent. Shortly thereafter, Watson and Crick saw a copy of Franklin's report of her research over the past year. Crick recognized both the strength and the implications of her data, which suggested that DNA consisted of two chains running in opposite directions and that the phosphate groups were likely on the outside of the molecule. Following some trial and error, Watson and Crick completed a final version of the model (Figure 7.2) and published their results on April 25, 1953, along with papers by Wilkins and Franklin that contained X-ray crystallographic data supporting the predicted structure.

By this time Franklin had left Randall's laboratory. Franklin died from ovarian cancer in 1958, at the age of 37. Four years later, in 1962, Watson, Crick, and Wilkins won the Nobel Prize for their work in elucidating the structure of

Figure 7.2: James Watson and Francis Crick, with their DNA double helix. A. Barrington Brown / Photo Researchers, Inc.

DNA; Franklin did not, because the Nobel Prize is not awarded posthumously. Until recently Franklin had not been given due credit for her role in contributing the essential data that allowed for Watson and Crick's constructing a model of DNA's structure. Much of this neglect occurred because she was a woman; this was not an uncommon reaction at the time to women scientists at King's College in Britain, who often held roles inferior to those accorded to men. Women also faced social obstacles. Franklin, for example, could not frequent the university's dining halls or pubs with her male colleagues. This social atmosphere was in stark contrast to the more egalitarian French culture to which Franklin had become accustomed while studying x-ray crystallography in Paris. But recently historians have increasingly understood that Franklin's role was crucial, and the importance of her empirical data is now widely acknowledged.

DNA REPLICATION

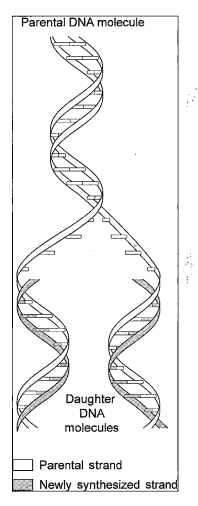

In their 1953 publication of the three-dimensional structure of the DNA molecule, Watson and Crick pointed out an important property of their proposed structure:

> . . . if only specific pairs of bases can be formed, it follows that if the sequence of bases on one chain is given, then the sequence on the other chain is automatically determined. . . . It has not escaped our notice that the specific pairing we have postulated immediately suggests a possible copying mechanism for the genetic material. (Watson and Crick, p. 738)

The mechanism of DNA replication envisaged by Watson and Crick was described in a later article: a double-stranded DNA strand separates into two single strands, and each strand acts as a template for the creation of another strand. Since bases pair in only one possible way (A with T, and G with C), the newly constructed strands have a predictable base sequence, complementary to the sequence of its template and identical to the sequence of the original strand that it replaces. The mechanism is described as *semi-conservative replication,* because the two new DNA molecules each contain one old and one new strand (Figure 7.3).

Figure 7.3: Semi-conservative replication. Illustration by Jeff Dixon.

In 1957, semi-conservative replication was experimentally confirmed by Matthew Meselson and Franklin Stahl at Caltech. Meselson and Stahl grew bacteria in a flask containing a heavy isotope of nitrogen, transferred these bacteria to a flask containing the normal (light isotope) form of nitrogen, and left the bacteria to grow for 20 minutes, enough time for about one round of DNA replication. If replication is semi-conservative, the newly replicated DNA molecule should contain one light and one heavy chain, which is what Meselson and Stahl observed. At about this same time, Arthur Kornberg at the University of Illinois discovered an enzyme that he called DNA polymerase, which played a central role in DNA replication.

HOW DNA FUNCTIONS AS A GENE

Thus DNA replication was predicted by Watson and Crick from its structure and confirmed experimentally in a relative quick amount of time. But an additional and more complex question was yet to be answered: how does DNA function as a gene? Organisms are constructed out of proteins, and, as discussed in Chapter 6, proteins were extensively studied by this time. The larger question of how DNA functions as a gene was therefore reduced to the more specific but still difficult question of how DNA determines protein synthesis. It was well known that proteins were very large molecules consisting of smaller protein subunits called *amino acids,* strung together linearly to form a chain. Those who became interested in DNA suspected two things about how DNA made proteins. First, the base-pair sequence order was the key. DNA, it was believed, somehow contained in the order of its base pairs the information necessary to determine the amino acid sequence of specific proteins for which DNA encoded. Second, *RNA* was involved.

RNA, or *ribonucleic acid,* is a nucleic acid similar to DNA, but only single-stranded and containing a uracil (U) base in place of the thymine (T) base present in DNA. RNA had been known since the mid-1930s to be present in the cell in large quantities during protein production. It was also known by the early 1950s that protein production occurred on the surface of convoluted membranes inside the cell called the *endoplasmic reticulum.* In 1953, George Palade discovered what he called *microsomes,* small protein/RNA complexes located on the endoplasmic reticulum. Protein production occurred here at these microsome sites.

In 1953, George Gamow, a Russian-born astronomer and physicist, read Watson and Crick's paper and wrote them a letter describing a theory he had for how their DNA structure might code for proteins. Gamow suggested that the protein subunits, or amino acids, migrated into the nucleus and somehow fit themselves into grooves on the external surface of the DNA molecules. Gamow imagined that the shape of these grooves was determined by the order of the DNA base sequences. He reasoned that since the shapes of the bases differ slightly, their order might affect groove shape. He thought that each possible groove might be shaped in such a way that it could accept only one possible amino acid.

Watson and Crick disagreed with Gamow's theory, and history has shown that they were right to do so: protein synthesis does not occur by the migration of amino acids into Gamow holes. Nonetheless, Watson and Crick enjoyed Gamow's company, and the three men struck up a friendship. In March 1954, Gamow formed the *RNA Tie Club* with Watson, Crick, and a few molecular biologists and others interested in the problem of protein synthesis. The RNA Tie Club was a half-joking, tongue-in-cheek social club, but in fact it became an important focal point of DNA research, with an increasingly large roster of members interested in how DNA coded for proteins and how RNA was involved. The work of the group was very theoretical and mathematical, attempting to provide theoretical interpretations of experimental work done by others. Gamow remained the driving force behind the Club.

But it was Francis Crick who became a central figure in the search for the mechanism of protein synthesis. In mid-1954, Crick wrote a letter to the RNA tie club members, hypothesizing that a three-base-pair combination, or a *triplet,* encoded one amino acid. Crick reasoned that two base pairs would not be enough: there are only 16 possible combinations of these, an insufficient number to encode all known amino acids, of which there were known to be at least twenty. But a three-base-pair code gives 64 possible combinations, more than enough to encode all amino acids. Crick suggested that the extra combinations of triplets could code for other aspects of protein production such as initiating or stopping the process, or there could be more than one code per amino acid as a backup mechanism for the process.

About half a year later, in early 1955, Crick wrote another letter to the club members, suggesting the existence of what he called an *adaptor molecule,* a molecule that would function to carry amino acids to the microsomes for protein production. Crick suggested that one end of this molecule attached to an amino acid, while the other end contained a three-base-pair sequence matching the amino acid's base-pair code. Not long after this, a molecule was discovered that seemed to have the properties that Crick had theorized about. By 1962, these properties were confirmed experimentally, and the molecule was named *transfer RNA* (Figure 7.4).

In the late 1950s, Crick and others assumed that Crick's adaptor molecule carried amino acids to the microsomes and that the microsomes contained a replica of the DNA code and thus acted as a template for protein synthesis. But, by 1958, experiments

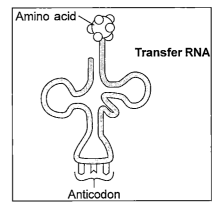

Figure 7.4: Transfer RNA. Illustration by Jeff Dixon.

by Francois Jacob and Jacques Monod challenged this assumption. Using bacterial cells in which they could trigger protein production under very controlled experimental conditions, Jacob and Monod showed that protein production begins very quickly after triggering and could also be stopped very quickly. This short time interval seemed to be insufficient for the cell to build and break down microsomes (which were pretty complicated structures) from scratch. Other experiments suggested that microsomes (called *ribosomes* by 1958) were very stable and existed as permanent structures in the cell.

If this was true—if ribosomes were permanent structures—then they could not contain the code for protein synthesis: there had to exist another molecule that somehow transferred the DNA code to the ribosomes. In March 1960, Sydney Brenner, a British molecular biologist and a member of the RNA Tie Club, wrote a letter to club members reviewing various lines of research completed throughout the 1950s. He noted that some people had suggested a third type of RNA that was assembled and broken down quite quickly, that seemed to correspond to DNA base sequence in many cases studied, and that appeared to be concentrated at the ribosomes during protein production. Previously this research had been overlooked by members of the RNA Tie Club. Brenner suggested that this third type of RNA carried DNA's genetic code from the nucleus to the ribosomes in the cytoplasm and acted as the template for protein production. He called this genetic RNA, but the molecule was renamed *messenger RNA* by Jacob and Monod, and the name stuck.

By this time, therefore, it was understood that DNA was copied into RNA and that the RNA copy then migrated to the ribosomes where it was used for protein synthesis (Figure 7.5). Despite these advances, however, there was little progress in the mid- and late 1950s toward understanding the *genetic code:* how did DNA specify which amino acids were to be used during protein production? The solution to the problem of the genetic code came not from Crick or other central figures of the RNA Tie Club. Instead, it came from an outsider biochemist with no prior knowledge of the RNA Tie Club or its work. In August 1961, Marshall Nirenberg, a biochemist at the U.S. National Institutes of Health, spoke at a conference in the Soviet Union, describing his group's work on the code. While many members of the RNA Tie Club attended the conference itself, none attended Nirenberg's talk, since he was completely unknown by molecular biologists.

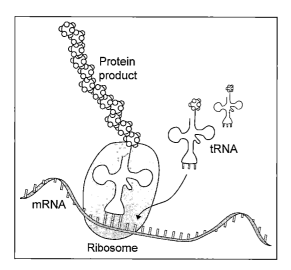

Figure 7.5: Protein synthesis at the ribosome. Illustration by Jeff Dixon.

Nirenberg had designed a way to make proteins outside the normal cell environment, using RNA as a template for synthesis. He and his team at the NIH had built an artificial messenger RNA containing only a string of uracil bases, and they used this artificial RNA to construct an amino acid chain that consequently contained only one type of amino acid, phenylanine. He concluded that phenylanine was encoded by a sequence of uracils (he did not entertain a guess as to how many uracils were required). Crick heard about Nirenberg's talk and realized that here was the beginning of a solution to the problem of the genetic code that had eluded the members of the RNA Tie Club. Crick arranged for Nirenberg to give the talk once again at the same conference, but this time in front of hundreds of molecular biologists. Everyone was electrified by the results. Molecular biologists, who had generally not even considered this sort of biochemical approach to the problem of the genetic code, quickly duplicated and extended Nirenberg's results. Crick himself confirmed that it was in fact three uracils that coded for phenylalanine, proving his prior theory that three bases code for one amino acid. By 1966, the entire genetic code was elucidated using Nirenberg's methodology (Figure 7.6).

		Second Letter			
	U	**C**	**A**	**G**	
U	UUU ⎤ Phe UUC ⎦ UUA ⎤ Leu UUG ⎦	UCU ⎤ UCC ⎥ Ser UCA ⎥ UCG ⎦	UAU ⎤ Tyr UAC ⎦ UAA Stop UAG Stop	UGU⎤ Cys UGC⎦ UGA Stop UGG Trp	U C A G
C	CUU ⎤ CUC ⎥ Leu CUA ⎥ CUG ⎦	CCU ⎤ CCC ⎥ Pro CCA ⎥ CCG ⎦	CAU⎤ His CAC⎦ CAA ⎤ Gln CAG⎦	CGU⎤ CGC ⎥ Arg CGA ⎥ CGG⎦	U C A G
A	AUU⎤ AUC ⎥ Ile AUA⎦ AUG Met	ACU⎤ ACC ⎥ Thr ACA ⎥ ACG⎦	AAU⎤ Asn AAC⎦ AAA⎤ Lys AAG⎦	AGU⎤ Ser AGC⎦ AGA⎤ Arg AGG⎦	U C A G
G	GUU⎤ GUC ⎥ Val GUA ⎥ GUG⎦	GCU⎤ GCC ⎥ Ala GCA ⎥ GCG⎦	GAU⎤ Asp GAC⎦ GAA ⎤ Glu GAG⎦	GGU⎤ GGC ⎥ Gly GGA ⎥ GGG⎦	U C A G

First Letter (left side) / Third Letter (right side)

Figure 7.6: The genetic code.

THE LAC OPERON AND REGULATION OF GENE EXPRESSION

Another important early discovery was the means by which gene activity was regulated. Genes of course do not produce proteins all the time; otherwise, cells would not be able to specialize in order to form specific body parts. Genes are turned on and off in different cells at different times, which allows an organism to grow, develop specialized structures, and carry out specific metabolic functions as needed, among other things. The means by which such regulation occurs was a big question following the publication of the Watson-Crick model.

A major discovery in the early 1960s elucidated some central methods of gene regulation. In 1961, Francois Jacob and Jacques Monod elucidated the regulation of a set of enzymes in *Escherischia coli,* a common gut bacterium. The enzymes function in the production of energy from lactose sugar. Specifically, they function in the digestion of lactose into the simpler sugars glucose and galactose. Jacob and Monod found that this set of genes produced its enzymes together, at the same time, and that a complex feedback system controlled when these enzymes were produced. They called such sets of regulated genes *operons,* and they called the specific lactose system the *lac operon* (Figure 7.7).

The lac operon consists of six different elements adjacent to each other on a bacterial DNA strand: three regions of DNA, called the *promoter,* the *operator,* and the *terminator,* and three genes called lacZ, lacY, and lacA. The lacZ

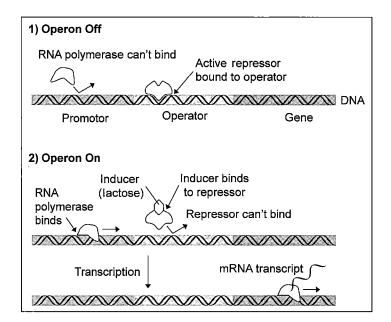

Figure 7.7: The lac operon. Illustration by Jeff Dixon.

gene encodes beta-galactosidase, an enzyme that breaks lactose into glucose and galactose. The lacY gene encodes a permease enzyme, which carries lactose into the bacterium. The lacA gene encodes transacetylase, an enzyme that destroys any toxins that the permease might accidentally carry into the bacterium.

These three genes are simultaneously triggered to produce large amounts of the proteins they encode (in other words, the genes are *expressed*) if a protein called RNA polymerase binds to the promoter. RNA polymerase functions to produce RNA from DNA (which, as we saw in the preceding chapter, is the first step in protein production from DNA). However, these genes are shut off and cannot produce their proteins (or, more accurately, they produce them in very small amounts) if a protein called the *repressor* binds to the operator. In this case, the binding of the repressor prevents RNA polymerase from binding to the promoter, which in turns prevents gene expression—or allows it to occur only in extremely small amounts.

This complex process is controlled by the presence or absence of lactose. If no lactose is present in the bacterium's surrounding environment, there is no need for the lac operon to produce the enzymes needed to digest it. In this situation, therefore, the repressor protein binds to the operator, which prevents the lac operon from producing its enzymes (by blocking the binding of RNA polymerase to the promoter). However, if lactose suddenly becomes available for the bacterium's use, a lactose by-product called *allolactose* binds to the repressor and changes the repressor's shape, preventing it from binding to the operator. The allolactose is produced by the breakdown of a very small amount of lactose, which occurs when the three genes of the lac operon are repressed but still expressing at extremely low levels. This binding of allolactose to the repressor allows the lac operon to activate; its three genes are then expressed at high levels, and the lactose is used by the bacterium for energy production (Figure 7.7).

CONCLUSION

While Avery's early experimental results suggested that the role of the gene was played by the DNA molecule rather than protein, protein remained the more popular alternative until the early 1950s. But Watson and Crick's compelling structural model, complete with suggestions about how DNA might replicate and code for protein production according to its base-pair sequence, produced a complete shift in focus within molecular biology to DNA. For the next 15 years, a growing community of molecular biologists sought the mechanism of protein synthesis and attempted to crack the genetic code—a task often compared at the time to wartime cryptographic efforts to crack the codes of enemy transmissions. Molecular biologists dominated this activity, with Francis Crick at the center of most of these efforts, and a great deal of theoretical and experimental research successfully outlined the process by which DNA produced RNA, which was then used as a template for protein synthesis. But it was the work of a group of biochemists at what was at the time a relatively unknown government facility, the National Institutes of Health, that finally cracked the code itself, using biochemical techniques of artificially synthesizing proteins in order to determine which base-pair triplets encoded which amino acids.

By the late 1960s, a great deal was known about how genes functioned to produce proteins. A consequence of this increased understanding of the machinery of protein production was that many molecular biologists began to envision the possibility, inspired by Marshall Nirenberg's artificial system, that proteins could be modified and mass-produced artificially, using genetic manipulation techniques. This sort of research eventually resulted in the creation of *recombinant DNA technology,* also frequently called *genetic engineering,* a method to combine DNA from very different organisms. This technology also became the foundation for a major new industry—biotechnology, the subject of the next chapter.

8

MANIPULATING DNA

INTRODUCTION

Following the discovery of the structure of DNA, researchers over the next several decades achieved spectacular successes in the analysis and experimental manipulation of the DNA molecule. By the late 1960s to early 1970s, focused research based on the Watson-Crick model produced dramatic results. As described in Chapter 6, the genetic code was deciphered, the cellular machinery involved in DNA replication and protein synthesis were increasingly understood, and the nature of gene regulation was elucidated. In the process, molecular biologists and biochemists became increasingly confident about the possibility of not just studying the gene but manipulating it in a variety of ways for a variety of purposes.

By the early 1960s, biologists saw the potential for increasing advances in medical knowledge and practice, and generally for a technological approach to improving social well-being, as a major reason for exploring the possibilities of altering our genes. By the mid-1970s, these possibilities and goals reached their peak with the creation of *recombinant DNA technology*, a method that allowed researchers to combine DNA from organisms radically separated from each other evolutionarily. The immediate purpose was medical and commercial: scientists could put gene coding for biological drugs such as insulin, interferon, and human growth hormone into bacterial cells, which would then divide rapidly, producing many copies of the gene and secreting large volumes of the drugs for medical use. These possibilities sparked the formation of a very successful biotechnology industry, consisting of new small companies, founded by scientists and funded by investors who saw their commercial potential. These companies either marketed their own products or partnered with established pharmaceutical companies to develop and market their products; in the process, many molecular biologists became millionaires. Meanwhile, both the new technologies and the new industry it spawned sparked concerns, both within the scientific community and among the general public, about the potential for creating new human and environmental hazards, the possibility of a "new eugenics," and the commercialization of university research.

RECOMBINANT DNA TECHNOLOGY

Recombinant DNA (rDNA) technology refers to a collection of techniques that allow for experimental manipulation of DNA. The techniques have had enormous effects on molecular biological research and are also the foundation of the modern biotechnology industry. One of the most common uses of these techniques is to combine DNA from radically different organisms, such as humans and bacteria. The combining of DNA from radically different forms of life does not happen in nature and is unprecedented in the history of biology.

In late 1971 and early 1972, Peter Lobban, a graduate student, and Paul Berg, departmental Chair, both at Stanford University's Department of Biochemistry, independently combined DNA artificially from two different viruses. Both used the same method: double-stranded DNA loops from two viruses were cleaved at one location, in order to linearize the loops; a short string of complementary base pairs (adenine [A] to one, and thymine [T] to the other) was then added to the two strands. The strands were then mixed under specific conditions that caused them to attach to each other and form a circle. This new hybrid, or recombinant, DNA molecule produced by Berg and Lobban's technique did not function. The procedure damaged the DNA, and the addition of adenine and thymine prevented it from properly producing protein.

Not long after these first experiments, Herbert Boyer, at the University of California-San Francisco, and Stanley Cohen, at Stanford University, devel-

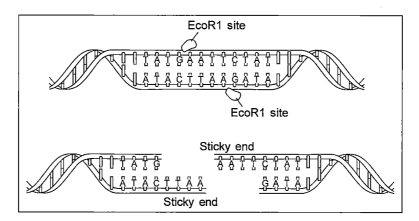

Figure 8.1: *EcoR1* cleaving DNA. Illustration by Jeff Dixon.

oped techniques to hybridize two DNA strands without requiring the addition of adenine and thymine. In 1970, Boyer discovered restriction enzymes, a class of enzymes in bacteria that cut DNA at very specific locations by recognizing specific base sequences. The natural function of restriction enzymes is to cut foreign, pathogenic DNA, a defense mechanism for bacterial cells against infectious (typically viral) agents. By the early 1970s Boyer was working extensively with *EcoR1*, an enzyme his laboratory isolated and characterized, and found that it cut at an angle, leaving single-stranded base pairs, dubbed *sticky ends*, at the point of cleavage (Figure 8.1).

Boyer found that two pieces of DNA stick to each other, or *anneal*, if they are each cut with the same restriction enzyme. This activity of restriction enzymes offered the potential for a much simpler method for combining two pieces of DNA artificially than was possible with Lobban and Berg's procedure of chemically cutting DNA and adding adenine and thymine to the ends. From the fall of 1972 to late 1973, Boyer and Cohen attempted to use *EcoR1* to anneal DNA from two different species. They first attempted to use viral DNA but were unsuccessful, because *EcoR1* cut the circular viral DNA molecules into multiple pieces and thus rendered it useless for the experiment.

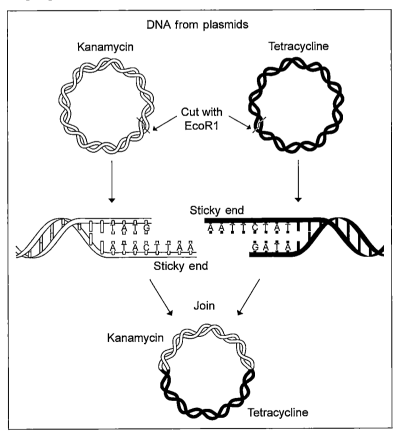

Figure 8.2: Recombinant DNA technology. Illustration by Jeff Dixon.

Eventually they tried plasmids, short, circular DNA strands that occurred naturally in bacteria. By the early 1970s, quite a few of these had been isolated and used. By trial and error, the group found a plasmid, pSC101, that was cut only once by *EcoR1* and, as a result, was linearized. The plasmid also contained a gene coding for a protein that gave the bacteria resistance to an antibiotic, called kanamycin. The group cut this plasmid with *EcoR1* and added to it a gene, also containing *EcoR1* sticky ends but coding for resistance to another antibiotic, tetracycline. When the two DNA strands were combined and put in bacteria, the bacterial cells were resistant to both antibiotics, indicating that they did in fact seem to contain both antibiotic genes and, therefore, the recombinant plasmid (Figure 8.2).

In late 1973, Boyer's and Cohen's research groups inserted frog DNA into bacteria using the same techniques. The bacteria did not produce the protein encoded by the inserted DNA, but they did produce RNA. It was assumed that protein production was imminent. From 1974 to 1977, rDNA techniques spread and were fine-tuned, but in fact there was little success with mammalian protein production. Mammalian genes were discovered to be significantly more complex than viral and bacterial genes, and the bacterial cell could not process mammalian DNA in a way required for protein production. But finally, in 1977, the group successfully produced somatostatin, a protein found in the mammalian brain, using rDNA technology. Human protein production had thus been demonstrated, and the technology underwent rapid development for the next few years.

RECOMBINANT DNA AND SOCIAL CONTROVERSY

Both scientists and the general public reacted to the novelty of DNA manipulation, expressing various levels of excitement, awe, and concern about the implications of this new power to control and alter the hereditary material. In response to concerns publicly voiced by Paul Berg and others, in 1973 the U.S. National Academy of Sciences formed a panel, chaired by Berg, to discuss the implications of rDNA research; in 1974, the panel recommended a voluntary moratorium on research until protocols could be developed under the auspices of the National Institutes of Health.

The research panel's recommendations, and Cohen's 1973 experiments with inserting frog DNA into bacteria, stimulated an increased public focus on rDNA technology; the insertion of frog DNA into bacteria especially emphasized the potential power of the new technology to break the species barrier. Media attention in turn prompted further public discussion of rDNA by biologists, who became concerned that the public response would neglect the potential benefits of the technology and who consequently put "potential" or "speculative" hazards, such as contamination, infection and rDNA-based warfare, in the context of the enormous scientific and industrial benefits of the technology, describing the technology as precipitating a medical, social, and economic revolution. The media similarly discussed rDNA technology as a "double-edged sword," with the potential for both tremendous benefits and potential harm.

In 1974, two committees were formed to examine the potential social implications of rDNA technology. One reported to the U.S. National Academy of Sciences and was chaired by Paul Berg (the Berg Committee), and the other reported to U.K. Advisory Board to the Research Councils, chaired by (Lord) Eric Ashby (the Ashby Committee). A year later, in February 1975, an international conference, the International Congress on Recombinant DNA Molecules, was organized by the Berg and the Ashby committees and held at Asilomar, California. While the "Asilomar conference" has frequently been described as serving to discuss the potential hazards of rDNA research, in fact only a portion of the conference was devoted to such hazards; the majority of the conference involved discussions of the contemporary and potential research opportunities using the technology.

In 1976, guidelines for using rDNA were established by the U.S. National Institutes of Health (NIH) Recombinant DNA Advisory Committee (RAC), based on recommendations from the Asilomar Conference. These were mandatory for NIH grantees but voluntary, although strongly encouraged, for industry. The content and the process of formation of these guidelines were quite controversial, precipitating significant debate within the research community with respect to their adequacy; the private sector benefited from closer access to the NIH through the voluntary compliance program, and guidelines were subsequently reduced, allowing the technology to be developed rapidly. The RAC process, controlled and administered by scientists themselves, served to legitimate and encourage, not to slow, the growth of industrial biotechnology.

But shortly following the release of the RAC guidelines, public debate erupted. Media coverage emphasized such concerns as the creation of strange new organisms, the danger of science-based social control and eugenics, and the perception of an overreliance on a "genetic fix" to social problems. Cambridge, Massachusetts, one of the cities where rDNA research was done, introduced a short-term moratorium, and Congress began considering legislation and regulation that would extend far beyond RAC's recommendations. In response to this perceived interference, many scientists became increasingly involved in lobbying against regulation. Whereas in the lead-up to the development of RAC guidelines the scientific community had expressed concerns about rDNA technology, in the aftermath of public controversy it more typically argued that the hazards of rDNA were greatly exaggerated, while the benefits were enormous. This change in strategy was not based on new evidence but was rather a consequence of mounting concerns about public interference in scientific activity, which was seen as a threat to the autonomy, and therefore the proper efficient functioning, of the enterprise of science itself.

Three meetings were held by scientists to organize a collective response to growing public and governmental criticism: in August 1976, at the National

Institutes of Health in Bethesda, Maryland; in June 1977, in Falmouth, Massachusetts; and in January 1978, in Ascot, in the United Kingdom. A distinct political agenda emerged during these meetings with respect to what constitutes
risk for rDNA use. While the Ascot meeting involved some significant levels
of debate about a wide range of issues, such as the potential for a variety of
recombinant organisms to be hazards, the potential for the transfer of recombinant DNA to other organisms, or the risks of various inserted genes, generally
the agenda and the subsequent tone of debate at the meetings served to down-
play risk and to devise a strategy to counter those critical of the consequences
of rDNA research. Discussion of risk was reduced to a technical discussion of
the possibility of serious outbreak of the organism that, at the time, dominated
rDNA research: recombinant *E. coli* K12. By this point, a recombinant K12
outbreak was already known to be highly unlikely; the reduction of risk to this
one issue, according to some historians, is best explained as an attempt by the
participants to capture ownership of risk discussion. Subsequent reports of
all meetings downplayed further the existence of debate within the molecular
biology community and suggested a relatively unanimous belief among biologists that the risk was minimal, while the benefits were enormous. This activity
successfully removed the threat of external, congressional legislative regulation of rDNA research, and the original NIH guidelines were markedly relaxed
and subsequently removed.

PRACTICAL ROLES OF RECOMBINANT DNA TECHNOLOGY

Recombinant DNA revolutionized the academic field of molecular biology, providing a means to easily isolate, grow, and manipulate DNA of interest in order to study gene structure and function. Its proponents within the
molecular biology community saw it as an invaluable tool that would shed
light on many scientific questions. But another major reason for interest
in the technologies was that it would provide many medical benefits. The
development of rDNA technology was seen as constituting a revolution in
medicine, and, as described previously, this view was used to justify its
continued in the face of concerns about hazards. A 1975 article discussing
the risks and benefits of the technology, for example, took it for granted that
medicine would eventually be broadly revolutionized by the new technology and warned that those who seek to impede the research must weigh any
potential risks in this context. These many potential medical benefits of
recombinant DNA research were by this time widely agreed upon and used
to medically and morally justify continued development of the research.

Yet, increasingly, another clear reason for the popularity of the technology
was its commercial possibilities. In fact, in the mid-1970s, recombinant DNA
technology gave birth to a new industry, called *biotechnology*. It is important
to acknowledge that other technologies contribute to the new biotechnology, in

particular monoclonal antibody production, but also techniques for large-scale biological processing, cell-culturing techniques, advances in enzymology, and cell-fusion techniques. But rDNA technology has been widely acknowledged as the primary backbone of modern biotechnology.

THE EARLY HISTORY OF BIOTECHNOLOGY

Today's "new biotechnology" is in fact an extension of an earlier industrial tradition of understanding the biology of yeast fermentation for industrial purposes, and in particular brewing, which was a major established industry by the late nineteenth century in England, Germany, and the United States and a major contributor to the economies of these countries. An early research tradition called *zymotechnology* became established in these countries to study the biology and fermentation capacities of yeast and other microorganisms, in a conscious attempt to move beyond traditional craft-based fermentation techniques and to link the scientific understanding of yeast biology to the practical goals of industrial fermentation. Major nineteenth-century international zymotechnic institutes and research programs dedicated themselves to the biology of yeast fermentation, and especially brewing. The brewing of beer is a complex process, requiring an understanding of the primary ingredients (hops, barley, yeast, and water) and the complexities of their interactions. Consistent high-quality brewing was difficult, and input from several scientific and engineering disciplines was central to a move from a craft-based tradition to large-scale industrial production.

During World War I, the fermentation process was used to produce an increasingly large number of valuable products, and zymotechnology became an increasingly broad-based research program with major industrial applications. In 1917, Karl Ereky, a Hungarian engineer, coined a new term, *biotechnology*, to encompass these and other expanded uses of organisms for industrial production. In the face of food shortages during World War I, Erecky opened a pig-fattening center—an early factory farm—and explicitly referred to his pigs as biotechnological machines (*biotechnologische Arbeitsmaschinen*) that converted inputs of sugar beets into outputs of meat. The term *biotechnology* quickly spread, particularly in the brewing industry, but also in industries devoted to drug and chemical production. By the 1960s, a strong tradition of industrial biotechnology existed, based on various chemical and microbiological processes and spanning a broad spectrum of industries.

THE "NEW BIOTECHNOLOGY"

The phrases *old biotechnology* and *new biotechnology* are sometimes used to distinguish between the early biotechnology tradition and the biotechnology that has developed since the creation of recombinant DNA technologies in the early 1970s. The distinction is somewhat misleading; the two traditions have important continuities that help explain why recombinant DNA became commercially interesting so rapidly, and, especially in the early years of rDNA technology, both were often referred to as simply *biotechnology*. But at the same time, there are crucial differences between the old and the new biotechnology. The new biotechnology involved a fundamental change in *how* organisms were used for industrial purposes, allowing unprecedented modifications of organisms unlimited by species barriers or the norms of reproductive biology. Additionally, the early successes of rDNA technology caused biotechnology as a whole to grow rapidly in prominence as a focus of industrial development.

Figure 8.3: Herbert Boyer, co-founder of Genentech. © Douglas A. Lockard.

In the United States in particular, biotechnology became a major industry, and industrial concerns became much more a part of molecular biology than they ever had before. For example, the production of somatostatin by Herbert Boyer's group was completed in the context of Boyer having formed a new company, called Genentech, in 1976 (Figure 8.3). His success triggered a rapid period of commercial growth, as scientists formed new companies and collaborated with large pharmaceutical and other companies to bring products to market and as universities began to take an interest in commercial possibilities of their scientists' research.

This "explosive new industry," as the press often described it, was unusual in that it initially involved the creation of small companies by university-based molecular biologists. While it has always had an important practical element, molecular biology became much more commercialized following the development of recombinant DNA, and the boundaries between academic recombinant-DNA technology and commercial biotechnology became increasingly blurred.

BACTERIAL FACTORIES

Recombinant DNA's earliest conception was as a potentially powerful technique for large-scale production of commercially important pharmaceutical proteins, such as insulin, interferon, and human growth hormone. The use of bacteria as *factories* was a common metaphor used to describe this vision, which in fact predated development of the techniques themselves.

The earliest recombinant DNA experiments involved attempted produc- tion of commercially valuable pharmaceuticals in commercial settings or with commercial goals in mind. In 1977, Genentech produced somatotrophin using rDNA technology, and in 1978 it produced recombinant human insulin in col- laboration with the multinational corporation Eli Lilly. In 1980, scientists at Biogen Pharmaceuticals produced recombinant human interferon in associa- tion with Schering-Plough Corporation; the result was accompanied by a great frenzy of publicity, but in fact the interferon did not function. But these events triggered an enthusiastic media and commercial response, and in 1980 public stock offerings by the first biotechnology companies began to appear. Stock prices typically soared for these early companies, as investors became excited about the commercial possibilities of recombinant DNA.

GENETICALLY MODIFIED ANIMALS

In 1980, a patent was granted to Ananda Chakrabarty, a biochemist at General Electric, for a bacterium engineered to consume oil. The patent grant followed several years of controversy. Following an initial rejection, Chakrabarty took his case to the U.S. Supreme Court, which upheld the patent request. The contro- versy centered on whether the bacterium, and indeed any life form, constituted a novel, patentable invention. Chakrabarty's patent was the first to be granted for a modified organism other than plants, which were patentable since 1930 under a special congressional allowance (the 1930 Plant Patent Act). In a context of significant political and economic lobbying from biotechnology proponents, the Supreme Court overturned the original rejection, arguing that it was sufficient that a living organism be altered in order to constitute a novel invention.

The Chakrabarty decision set a precedent for allowing patenting of geneti- cally modified animals: the next and most notable such patent was the grant- ing, in 1988, of a patent for the *Oncomouse*™, created in 1983 by Philip Leder of Harvard University. The *Oncomouse*™ was genetically modified to be very susceptible to cancer because of the insertion of oncogenes into the mouse embryo. Oncogenes typically function in cell regulation; abnormal expression of these genes has been linked to the development of various cancers. The patent decision gave its holder sweeping control over the creation of animals engineered to be cancer-sensitive; the Oncomouse™ patent included intel- lectual property protection not just for the product but for any technique used to insert an oncogene into any animal.

GENETICALLY MODIFIED PLANTS

Because of the early application of rDNA to the production of somatostatin, interferon, and insulin, much of the initial industrial enthusiasm for rDNA technology focused on its applicability to pharmaceutical production. But other applications were not long in coming to the attention of industry and research groups; the most important of these was in agriculture.

One early example of agricultural uses was the production of chymosin, or rennet, using rDNA technology. Chymosin is an enzyme traditionally extracted from calf stomachs that coagulates milk and is used in cheese production. A variety of research programs in the mid-1980s had demonstrated recombinant chymosin expression by inserting a calf-derived chymosin gene in bacterial, yeast, and fungal hosts, using techniques identical or similar to those of Cohen and Boyer's rDNA technology. The most successful of these programs was done at an American company called Genencorp, which used species of the *Aspergillus* genus, a filamentous fungus, to produce large relatively large quantities of chymosin. By 1990, Genencorp was marketing this chymosin, and today most chymosin used in cheese production is obtained using very similar recombinant DNA processes.

In the mid-1980s, the Monsanto Corporation, in St. Louis, Missouri, a major international chemical firm with, at that time, a growing interest in agricultural biotechnologies, was producing bovine growth hormone (BGH), also called bovine somatotrophic hormone (BST), using recombinant DNA methods to insert the gene into bacteria. BGH injection into dairy cows causes them to produce about 20 percent more milk than they do naturally. Recombinant DNA-based BGH production was an early target of antibiotechnology protest, and, despite its early development, it was not until the mid 1990s that Monsanto obtained approval from the U.S. Food and Drug Administration to sell the product in the United States; the product entered the market in 1994.

The production of recombinant chymosin and bovine growth hormone is in fact no different from the rDNA-based production of pharmaceuticals. Organisms are once again used as "factories" to produce a product that, in its final market form, is identical to the original. The production process, rather than the product itself, incorporates rDNA technology. But, by the late 1970s, a different sort of genetic engineering program began to take shape; a variety of research programs, pursued by academic biologists alone or in partnership with start-up companies or established firms, were experimenting with using rDNA technologies to alter the genomes of several commercially important plants, with the goals of improving crop production. Production of these genetically modified (GM) plants is similar in principle to the use of bacteria to produce proteins of value: foreign genes are inserted into cells, but in this case the modified cell, rather than solely the foreign gene, is the product of interest.

Inserting genes into plant cells is a more complicated process than using the cells of microorganisms. In particular, the cell walls of plant cells proved to be a barrier to most artificial DNA delivery systems. In the early 1970s, several research groups, one led by Mary-Dell Chilton at the University of Washington in Seattle, another led by Marc van Montagu and Jeff Schell at the Free University of Ghent in Belgium, and a third at the University of Leiden in the Netherlands led by Robert Schilperoot, discovered that the bacterial plant pathogen, *Agrobacterium,* known since the early twentieth century, had the natural ability to incorporate some of its own DNA into plant genomes and that the genes responsible for this ability were located on a large plasmid. Later describing the microorganism as a "natural genetic engineer," a phrase used repeatedly since its discovery, the same groups speculated that this function of *Agrobacterium* might be amenable to delivering novel genes to plants and began research programs attempting to do so.

In 1980, Monsanto hired as consultants several of these researchers—Chilton, who had moved to Washington University in Indiana, and Montagu and Schell, in Belgium. The relationships lasted for about a year, during which time, in exchange for research funding, the researchers provided Monsanto with their data and with recombinant DNA constructs, created using *Agrobacterium* and containing potential genes to be transferred. But by mid- to late 1981, these research groups had severed their ties with Monsanto and continued *Agrobacterium* research on their own.

Just over a year later, in a race to establish priority and exclusive rights to the products of this research, Monsanto, Chilton, and Schell and Montagu each had successfully transferred a gene from *Agrobacterium* to tomato plants and, within days of each other, had applied for competing patents on their work: Monsanto and the Belgian group in Europe, and Chilton and Monsanto in the United States. The Schell and Montagu patent application was successful in Europe, having been received by the patent office just before Monsanto's. Unlike the European patent procedure, however, American patents are granted according to who first invented the technique or product; it was not until 2005 that the Chilton and Monsanto patents were resolved at the U.S. Patent Office, in a complex settlement that gave both parties royalty-free access to the technology.

Agrobacterium cannot be used to insert genes into all plants; the reason for this is unclear. The most common alternative to *Agrobacterium*-mediated gene insertion is called *biolistics,* a technique first invented in 1984 by John Sanford at Cornell University. Often called the *gene gun* technique, biolistics involves coating tungsten particles with DNA and shooting them at high velocity into recipient cells. In collaboration with researchers at Pioneer Hi-Bred, one of the largest agricultural seed companies in the world, Sanford successfully used this technique in 1987 to insert a test gene into corn plant cells. Sandford began the manufacture and sale of his gene guns soon thereafter. By 1990, several groups were reporting success in adding genes to corn using the gene gun.

During the 1980s, a variety of research programs attempted to create commercially valuable products using *Agrobacterium*-mediated rDNA technology. Two of the earliest lines of research involved pest control, and while a great number of researchers and several companies were involved in these researches, the Monsanto Corporation was again a major player, investing an enormous amount of energy and resources into being the first to create patentable products.

One of these lines of research involved inserting into crops a gene from a bacterium called *Bacillus thuringiensis,* which produces a protein that kills insects that feed on the plant. Different strains of *B. thuringiensis* affected different insects, so there was significant commercial potential for creating so-called "*Bt* plants" with resistance to a variety of insects. Many groups pushed to produce this technology, but Monsanto was a leader, creating in-house research programs and hiring academic researchers as consultants. The other line of research involved creating plants with resistance to herbicides. Monsanto sold a broad-spectrum herbicide, called glyphosate or Roundup™, which functioned by destroying an enzyme crucial to plant and bacterial amino acid production. Glyphosate was a quickly degrading, weak-acting herbicide capable of killing almost all plant life and had become one of Monsanto's top products, marketed for situations in which all vegetation was intended to be eliminated. When scientists from the company found bacteria that seemed resistant to this herbicide, Monsanto and other groups spent most of the 1980s attempting to isolate the genetic source of this resistance and to insert the gene into plants, with the hope of being able to expand the glyphosate market by allowing the spraying of glyphosate-resistant (GR) agricultural crops to remove weed contamination.

By the mid- to late 1990s, genetically modified plants began to make it to the market. But the first agricultural product created was not *Bt* or GR plants; these did not see market approval until 1995 for *Bt* potatoes, and 1996 for GR soybeans. In 1988, two separate groups, a U.K. group at the University of Nottingham's Agricultural Science department, in collaboration with a U.K. company called ICI Seeds, and scientists at a small U.S. start-up company called Calgene, published results demonstrating the production of a tomato with diminished capacity for softening once it was ripe. Typically, tomatoes are harvested prior to ripening in order to allow time for shipping; otherwise the ripe tomato softens and is often damaged during the shipping process. The gene responsible for this softening process codes for an enzyme called polygalacturonase. The modified tomato was created using antisense technology, which involves using *Agrobacterium* to insert a short piece of complementary RNA that binds to this gene and prevents its functioning.

ICI, a major British chemical company (now called Zeneca), had filed for a U.S. patent for this technology in 1986, but in 1989 Calgene challenged it; it had been working in collaboration with Campbell's since the early 1980s and therefore claimed priority. The two companies eventually settled out of court: ICI obtained permission to sell its products as tomato derivatives, such as pasta sauces and ketchup, while Calgene would sell whole tomatoes. Calgene applied for FDA approval of its "Flavr Savr™" tomato in 1991 but did not obtain approval until 1994. But, even after approval, the modified tomato had little success; production costs and a relatively small difference in time to softening served to prevent Calgene's Flavr Savr™ from ever becoming a commercial success.

Since their introduction in the mid-1990s, genetically modified (GM) crops have been planted and grown in rapidly increasing numbers, with a land mass increase of at least 10 percent per year. By 2003, GM crops were being grown by approximately 7 million farmers in 18 different countries; most were grown in North America but there was increasing growth in developing countries. GM plants occupied approximately 70 million hectares (approximately 167 million acres) in 2003, a 40-fold increase from the first year of planting, 1996. As of 2003, approximately 25 percent of the four most commercially important crops grown in the world (corn, soy, cotton, and canola) were genetically modified. The global value of these crops was estimated in 2003 to be about US$5 billion, increasing at a rate of about three-quarters of a billion dollars per year.

Most of these transgenic plants are GR soybeans, *Bt* corn, and *Bt* cotton, all developed and patented by Monsanto in the 1990s. In the early 1990s, Monsanto sold permanent rights to produce GR soybeans to two other companies, Pioneer Hi-Bred International, Inc. (now owned by DuPont) and Novartis. A similar situation exists for the other crops; Monsanto holds the patent and has sold permanent rights of production to several other companies. Global transgenic crop production today is thus very much an *oligopolistic* industry, where a handful of very large multinational corporations dominate the market. These transgenic crops are typically used in food processing, and by most estimates approximately 60 percent of globally sold processed foods contain genetically modified ingredients.

BIOTECHNOLOGY AND STATE POLICY

The commercial development of biotechnology was paralleled by a growing state involvement in creating a favorable macroeconomic climate for the field and in encouraging commercialization of university research. The state began to more heavily support science research in the wake of World War II and the rise of state science policy programs. During and immediately following World War II, state science funding was generous and broad-reaching, but, as described later, by the late 1960s and early 1970s funding came to require a more practical justification, and governments increasingly engaged in planning exercises to maximize investments in science.

An important example of state intervention to encourage genetic and biomedical research was the declaration of "War on Cancer" in 1971 by then-President Richard Nixon, who promised a cure for cancer within five years, in time for the American bicentennial in 1976. The War on Cancer began with the creation of the 1971 National Cancer Act, which provided a crucial opportunity for the growth of cancer research. Despite a program of massive funding of cancer research since then, most commentators agree that, with the exception of childhood cancers, which accounted for approximately 0.2 percent of cancer deaths in 1993, little if any substantial progress has been made in reducing the overall death toll caused by cancer.

Although the War on Cancer has so far failed to affect cancer as one of the leading causes of death, it was very successful in another way: it resulted in a massive investment in biomedical research, especially in support of the emerging developments in DNA manipulation and its application in industry. The several billions of dollars of federal financial investment in biomedical research in the 1970s to fight the War on Cancer was pivotal to the development in that time period of recombinant DNA technology, DNA sequencing technology, and other advances that provided the technological foundations of the new biotechnology industry.

State interest in biotechnology as an important industrial strategy was related to broader moves by governments to encourage commercializable research in a number of high-technology fields, of which biotechnology was only one. The decline in the 1960s of traditional metal, automobile, and textile industries, a period of economic recession in the 1970s, and the rise of research-intensive high-technology industries such as computers, chemicals, and pharmaceuticals brought about a corresponding political climate that encouraged low regulation, heavy government investment in industry, and increased ties between universities and industry. It became much easier for university researchers in various fields, not only biotechnology, to receive government funds if they worked with industry, especially following the election of conservative leaders such as Ronald Reagan in the United States in 1980 and Margaret Thatcher in Britain in 1979. A variety of incentives were developed in the early 1980s to encourage universities and industries to commercialize the products of university research, in particular the 1980 Patent and Trademark Amendment Act (or Bayh-Dole Act) in the U.S., which gave universities the rights to take out patents on federally funded research. While this phenomenon was broad, biotechnology was central to this new focus on innovation policy. It was a key high-technology industry that governments prioritized; patent incentives were especially helpful for encouraging pharmaceutical industry development; and academic-industry collaboration in biotechnology was especially pronounced. In biotechnology in particular, university scientists rapidly became involved with industry, starting biotechnology companies with venture capital funding, receiving increased funding from industry for university laboratories and research centers, and partnering with industry to develop specific research projects. Biotechnology became the driving force behind the increased relationships between universities and industry.

CONCLUSION

The Watson-Crick model of the structure of DNA was the beginning of a revolution in molecular biology. As the function of the molecule became increasingly understood in the 1960s through a series of now-famous discoveries, such as the means of DNA replication, the meaning of the genetic code, and the process of protein production, molecular biologists came to realize that with this knowledge came the power to actually manipulate the molecule, perhaps in ways that might offer benefits to humanity.

In the 1970s, these dreams became a reality with the creation of recombinant DNA technology, a method that allowed the transfer of genes between organisms from different species. Beginning with the growing in "bacterial factories" of biological molecules of medical importance and continuing with the production of genetically modified foods, an increasing number of uses of recombinant DNA technology were found in molecular biology laboratories.

Many of these achievements were inspired by medical promise, but they were also a consequence of the realization that the technology could produce commercially valuable products. A biotechnology industry grew to take advantage of this new market, consisting of small companies founded by the new breed of scientist entrepreneurs, many of whom became very financially successful in the process. In the midst of this activity, both scientists and the general public expressed a variety of concerns about the potential hazards of recombinant DNA and about the commercialization of university research.

In the next chapter we will discuss how the promises and the concerns of recombinant DNA technology were repeated in a major scientific initiative that grew out of these advances in molecular biology: the Human Genome Project, a major international effort to map all of the genes and sequence all of the DNA in humans and other animals.

Public and Private Science

Many commentators have expressed concern that the commercialization of molecular biology has interfered with the integrity of the field, creating a new culture of commodified, "private" knowledge in place of a traditional, communal, "public" knowledge. Knowledge became increasingly profitable, and both scientists and others expressed concern about increased secrecy and the hasty acceptance of premature conclusions.

While some historians have pointed out that this contrast is somewhat overstated, since science has historically often involved rivalries, secrecy, profit, and competition, and since a great deal of valuable knowledge has come out of research done in biotechnology industries (Genentech scientists, for example, have regularly published excellent academic research papers), nevertheless valid criticisms have been directed at the increasing connections between science and industry. The commercialization of academic molecular biology, and the state's role in facilitating this, can cause research programs to be focused on issues that are potentially profitable rather than socially important. Additionally, the benefits of some molecular biology research can be exaggerated in all the excitement generated about the biotechnology industry itself. Some commentators have argued that, for example, sometimes the best way to improve medical care for people may not be profitable, and biotechnology research thus may not correlate with social need. These debates are ongoing, which, on a positive note, is a good sign: healthy social debate about the role of biotechnology can help ensure that it is used for the good of everyone.

9

THE HUMAN GENOME PROJECT

INTRODUCTION

This chapter is a broad history of the Human Genome Project (HGP), an international concerted effort to discover the genes and to sequence the base pairs
of the deoxyribonucleic acid (DNA) of humans and other organisms. Many
technologies were prerequisites for the possibility of such an initiative, but
this is not a simple story of technological determinism; the HGP did not arise
solely from its technical possibility. Instead, a diverse group of people, with
many and at times conflicting interests, engaged in processes of negotiation
that led to the subsequent organization of the initiative. In other words, the
HGP has a social and political history, and it is this history, in addition to the
project's technological dimensions, that we are emphasizing here.

The Human Genome Project is generally thought of as a concrete and well-
defined event that created the field of genomics and its related fields (proteomics, metabolomics, and others)—that is, it created the set of data for the
many sorts of analyses that are said to constitute these fields. The general
argument is that the HGP is a first step, a data collection phase, that will produce (and has in fact produced) a mountain of data for subsequent analysis.

The HGP ultimately became a step-wise process, involving genetic and
physical mapping during the early stages of the project and DNA sequencing once the technology was sufficiently improved and more cost-effective.
Genetic mapping refers to the indirect determination of relative distances on
a chromosome between genes or other detectable DNA sequences by analyzing the frequency with which they are inherited together. This is an indirect
rather than an exact measure of physical distance, but it is a very common
and useful procedure. There are also various techniques collectively referred to as "physical mapping," which are used to determine actual physical

relationships between DNA sequences. The most common contemporary physical mapping techniques involve breaking the very long chromosomal DNA strands into smaller pieces, which are more suitable for analysis, and then determining the correct order of these sequences as they would appear on the original chromosome. Finally, DNA sequencing can be thought of as a physical map that is of the highest possible resolution; sequencing provides knowledge of the physical structure of the DNA at a resolution of individual bases. Each of these processes is explained in more detail in this chapter.

The Human Genome Project is typically capitalized, and it is often described as one international, concerted effort to analyze the human genome. However, it originated in the United States in the mid-1980s as a result of several independent initiatives; it spread to other countries largely as a result of actions by scientists from these countries who were involved in the creation of the American project, and it gained political momentum when other nations saw the potential for U.S. dominance of the field. Additionally, the project involves the analysis of the genomes of many organisms besides just human beings. Given this diversity, the Human Genome Project is often more fruitfully referred to as the "genome projects," the "genome initiatives," or simply as genomics. Nevertheless, the American and other international projects exhibit a centrality and a directed control that are useful to emphasize, and the goal of mapping and sequencing human DNA has consistently been paramount. These features are easily highlighted by referring to the various initiatives as the "Human Genome Project" or the "Human Genome Initiative." All of these terms and more are used in this chapter to highlight the diverse yet highly coordinated nature of the HGP.

I begin with a description of some of the major technologies that enabled the completion of the goals of the HGP. I follow with some of the earliest activities that were the first steps of the larger project. I then describe the initiation of the HGP by the U.S. Department of Energy and, later, by the National Institutes of Health and the culmination of these efforts to form a centralized, joint project that eventually took on international dimensions. In the process, I highlight the many dimensions of the HGP: social, political, and economic, as well as scientific and technological. Specifically, in this narrative I illustrate the way in which the HGP was woven into an existing political and economic climate that used the possibilities of biomedical intervention as a legitimator for the development of biotechnology as a commercial endeavor.

THE TECHNOLOGIES OF THE HUMAN GENOME PROJECT

By the mid-1980s, a variety of relatively recent technical advances suggested that an ambitious project such as a genome initiative was in fact becoming technically feasible. As discussed previously, in 1973, Stanley Cohen, at Stanford University, and Herbert Boyer, at the University of California in San Francisco, invented recombinant DNA (rDNA) technology, a technique for combining DNA from different organisms, often from radically different species. Rapid advances occurred in rDNA technology in the late 1970s, including splicing genes from higher organisms such as humans into bacterial plasmids. The most significant advance from a commercial perspective involved inducing bacteria to produce the proteins encoded by these spliced genes, which potentially allowed for the rapid production of large amounts of desired proteins. By the early 1980s, valuable proteins such as insulin, interferon, and human growth hormone were being expressed in recombinant bacteria, and there was commercial anticipation that these proteins would soon be produced on an industrial scale.

In 1975, Ed Southern, at Oxford University, invented the Southern Blot. The procedure relies on two other technologies: gel electrophoresis (Figure 9.1), invented in 1955 by Oliver Smithies at the Connaught Laboratories in Toronto, for separating molecules of different sizes on a gel using an electric current, and DNA probes, small strands of radioactive DNA with a known base sequence. DNA probes can be bound to DNA with a base-pair sequence that matches that of the probe, allowing the detection of very specific DNA sequences from a complex DNA mixture.

These technologies can be used to produce a *physical map,* a collection of DNA fragments arranged in their proper order. If a probe hybridizes to inserted DNA in two different clones, this suggests that the inserts contain overlapping DNA sequences (Figure 9.2). A series of overlapping clones can thus be identified and put in the order corresponding to the base sequence order of the original genome. A physical map is a useful tool when searching for genes; if a gene has been genetically mapped to a specific region of DNA (see later discussion), then a physical map of this region can be used to locate and physically isolate this gene.

In 1977, methods for deciphering a base sequence of DNA were developed in two separate locations: at Harvard University, by Walter Gilbert and Allan Maxam, and at Cambridge University, by Fred Sanger. These DNA sequencing processes were automated in the mid-1980s and marketed by Applied Biosystems, Du Pont, and other companies. The "Sanger method" (Figure 9.3) is most commonly used; it uses dideoxynucleotides (ddNTPs), which are similar to deoxynucleotides (dNTPs), the normal components of DNA, but lack an oxygen atom. There are four forms, corresponding to the four dNTP types found in DNA: adenine (A), thymine (T), guanine (G), and cytosine (C). When any of these is incorporated into a synthesizing DNA strand, DNA synthesis stops. Sequencing involves four separate DNA synthesis reactions, each containing one of the four ddNTPs and each using the original DNA strand as a template. DNA synthesis occurs in each reaction tube, and ends when ddNTP gets incorporated into the strand. If the reaction is allowed to continue long

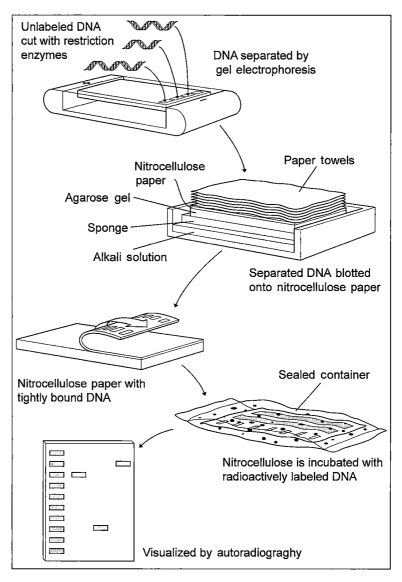

Unlabeled DNA
cut with restriction
enzymes

DNA separated by
gel electrophoresis

Paper towels

Nitrocellulose
paper

Agarose gel

Sponge

Alkali solution

Separated DNA blotted
onto nitrocellulose paper

Nitrocellulose paper with
tightly bound DNA

Sealed container

Nitrocellulose is incubated with
radioactively labeled DNA

Visualized by autoradiograghy

Figure 9.1: Gel electrophoresis. Illustration by Jeff Dixon.

enough, each of the four tubes will have a full assortment of DNA strands of varying lengths, all ending with the ddNTP present in the reaction tube. When these strands are separated by gel electrophoresis, the sequence of the original DNA molecule can be determined by simply reading the sequencing gel from bottom to top.

In 1985, Kary Mullis, then a technician with Cetus Corporation, invented the polymerase chain reaction (Figure 9.4), a process that could rapidly identify and replicate a segment of DNA, with a known sequence, out of a complex mixture of DNA. The technique was time-saving and simpler than the

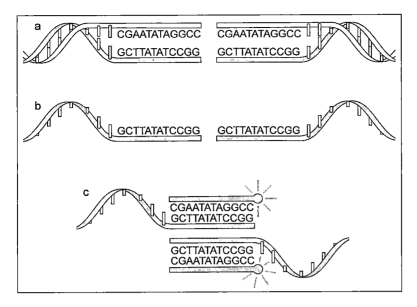

Figure 9.2: A DNA probe binding overlapping DNA sequences. Illustration by Jeff Dixon.

Southern Blot, and is used for a wide variety of applications, including genetic and physical mapping. PCR involves successive rounds of DNA synthesis of double-stranded DNA (dsDNA). The two strands are first separated and used as a template to synthesize another strand each, and these strands then form a new dsDNA. Both the original and the new dsDNA are then submitted to this same process. This cycle is performed anywhere from 20–40 times; the result is an exponential amplification of the DNA of interest (Figure 9.4).

A method for inserting human DNA into yeast cells using a yeast artificial chromosome (YAC) vector was invented in 1987 by Maynard Olson and David Burke, at Washington University. The new technique allowed for the insertion of much larger human DNA fragments than was previously possible, and it became very important in genomics. Physical maps constructed using YAC clones became the primary physical mapping tool used originally for the human genome project. However, these clones tend to be unstable; DNA inserts can rearrange spontaneously and therefore no longer represent the sequence as it existed in the donor genome. As an alternative to YACs, bacterial artificial chromosomes (BACs) were developed in 1992 at the California Institute of Technology; these were more stable and still allowed the insertion of relatively large inserts.

As early as the 1930s, human genetic mapping was conceived as a means for locating the genes for human disease. During meiosis (the production of *gametes*, or sperm and egg cells), the two copies of each chromosome exchange small pieces of DNA by a process called genetic crossover, or genetic recombination.

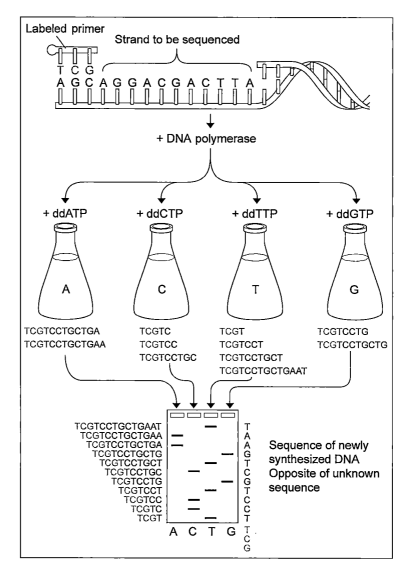

Figure 9.3: DNA sequencing—the Sanger method. Illustration by Jeff Dixon.

A crossover between two regions of DNA causes them to be separated and no longer co-inherited. If two DNA regions are co-inherited at a given frequency, these are said to be genetically linked, and the statistical frequency of co-inheritance is an indirect measure of their proximity. This "genetic distance" is measured in units called *Morgans,* named after Thomas Hunt Morgan, who discovered genetic linkage in fruit flies in about 1915. A Morgan is subdivided into 100 centiMorgans (cM); one cM corresponds to a 1 percent chance of recombination between two units of DNA. Thus, if two genes are described as being 5 cM apart, there is a 5 percent chance that a recombination event will

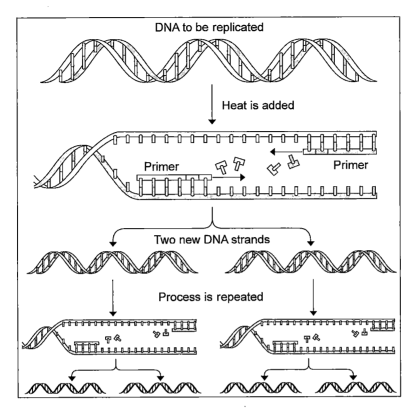

DNA to be replicated

Heat is added

Primer   Primer

Two new DNA strands

Process is repeated

Figure 9.4: The polymerase chain reaction (PCR). Illustration by Jeff Dixon.

occur between them, and the two genes are co-inherited 95 percent of the time. On average, 1 cM corresponds to a physical distance of one million base pairs, or one "megabase," but this correspondence varies significantly in different regions of a genome.

The first genetic mapping in humans was performed by Julia Bell and J.B.S. Haldane and published in 1937. Bell and Haldane successfully detected genetic linkage of the genes for hemophilia and a form of color-blindness, both of which were known to be located somewhere on the X chromosome. However, little work was done in human genetic mapping until the late 1960s. In 1967, Mary Weiss and Howard Green developed a technique called somatic cell hybridization, which produces hybrid cells containing mouse and human chromosomes. In 1971, Torbjorn Caspersson, at the Karolinska Institute, in Sweden, published a technique that produced consistent banding patterns on different chromosomes when they are stained with a dye. Using these two techniques, genes could be physically mapped to specific chromosomal regions—staining could determine which human chromosomes were in which cell hybrids, and genes could be localized to particular chromosomes according to the genes' presence or absence in specific cell hybrids. Additionally, the system functioned as a genetic mapping tool—during production of cell hybrids, the human chromosomes tend to break into fragments, and the cells keep only a portion of particular human chromosomes. The frequency with which two DNA regions tend to stay together during production of cell hybrids is proportional to their physical proximity. The concept is identical to traditional genetic mapping, but random chromosome breakage replaces genetic recombination as the process used to determine relative distances between DNA regions. Somatic cell hybrids were subsequently used to produce a variety of genetic and physical maps of the human genome.

Following the advent of rDNA technology, genetic mapping in humans accelerated rapidly. In 1978, Yuet Wai Kan, at the University of California in San Francisco, found evidence of linkage, in a collection of families, between the sickle-cell gene and a nearby piece of DNA sequence. The nearby region of the DNA, when cut with a restriction enzyme, produced differently sized fragments in healthy and sickle-cell individuals. This is called a restriction fragment length polymorphism (RFLP); these had been discovered and used earlier in 1978 by David Botstein at MIT and other colleagues in studies of yeast and virus genetics. Kan showed that this region of DNA was 13 cM from the gene for sickle-cell anemia, because the gene and one of the RFLP variants of the DNA sequence were inherited together 87 percent of the time $(100 - 87 = 13)$. Kan argued that RFLPs could be very useful for linkage analysis of human disease genes: they can be used as markers, and genes can be genetically mapped relative to these markers. Similar results were found by other research groups. In 1989, new markers were developed, detectable by PCR and called sequence-tagged sites (STS). These became very important for both physical and genetic mapping.

THE FIRST HUMAN GENOME MAP

David Botstein, from MIT, and Robert Davis, from the University of Utah, are generally credited with being the first, at a 1978 meeting at the University of Utah, to conceive of the creation of a dense genetic map made of RFLPs randomly dispersed throughout the human genome. In a 1980 paper, Botstein and Davis, together with their colleagues Raymond White, from the University of Massachusetts, and Mark Skolnick, from the University of Utah, suggested that a RFLP map could function as a framework to which disease genes could be genetically mapped and, subsequently, physically localized. The paper emphasizes the practical, medical uses of such an endeavor—specifically, polymorphic assays detectable in utero (during pregnancy) for use in "preventive medicine," or therapeutic abortion, and diagnostic markers for "predictive medicine," the predictive diagnosis of genetic disease, either for individuals or for couples at risk of having children with genetic diseases. In other words, Botstein's paper sought to demonstrate the practical uses of building a genetic map, rather than simply advocating for the scientific understanding of the human genome. This combination of basic and applied science is typical of much modern genetic research.

Botstein was a member of the scientific advisory board of Collaborative Research Inc., a company interested in the capacity for genomics to facilitate the discovery of new targets for drug development; Collaborative Research began construction of a map in 1983, and, by 1987, the group published the first genetic map of the human genome, using approximately 75 percent new RFLP data and 25 percent previously discovered, public RFLP data, many of which had been mapped by White and others at the University of Utah.

EARLY INTERESTS IN A SEQUENCING PROJECT

In 1985, Robert Sinsheimer, Chancellor of the Santa Cruz campus of the University of California, attempted to create a genome sequencing project centralized at Santa Cruz. Sinsheimer held a workshop in Santa Cruz with the hope of confirming the technical feasibility of obtaining a complete human genome sequence. However, the group concluded that a large-scale sequencing effort was in fact not feasible, given the current state of sequencing technologies. Instead, it recommended that efforts should concentrate on improving sequence technology and on sequencing regions of the genome that may be of immediate interest and relevance.

Sinsheimer was subsequently unsuccessful at raising funding for a sequencing initiative, but he did manage to pique the interest of other scientists. Beginning in 1986, Walter Gilbert, one of the conference participants (and the inventor of a method of genetic sequencing, for which he won a Nobel Prize in 1980), became an enthusiastic promoter of the project, within scientific circles and to the media. He tried several times to form companies centered around genome mapping and sequencing, although none were successful. He and others managed to obtain various research grants for large-scale sequencing efforts, and multiple independent genome projects, for viral, bacterial, animal, and human genomes, were created and began to grow.

Several years prior to Gilbert's activities, by 1984, the U.S. Department of Energy (DOE) was expressing an interest in the possibilities of a genome sequencing project. The DOE had a long history of interest in the mutagenic effects of radiation produced by both atomic weapons and industrial nuclear facilities: during the American postwar occupation of Japan, the DOE's precursor, the Atomic Energy Commission, funded the Atomic Bomb Casualty Commission (ABCC), which initiated a massive and controversial program of long-term study, without treatment, of the *hibakusha* ("those affected by the bomb"), Japanese survivors of the Hiroshima and Nagasaki bombings. Because the American government anticipated that the next large-scale war would likely involve nuclear weapons, the study aimed to observe radiation effects in anticipation of this event. A climate of fear existed following the

American nuclear attack on Japan, and much of the American population, including many scientists, envisioned the extinction of the human race as a consequence of the radiation released by the bombs. Some historians have also argued that, in order to maintain its nuclear industry, the U.S. government wished to allay public fears about nuclear research dangers. The *hibakusha* study was therefore closely tied to political, in addition to scientific, interests in the effects of radiation poisoning.

In 1975, the ABCC became the Radiation Effects Research Foundation (RERF). In a 1984 RERF report, one of the recommendations of the *hibakusha* study was that cell lines be produced from the survivors and that methods for direct analysis of their DNA be created. The RERF held a conference in December 1984, funded by the Department of Energy, to discuss DNA sequencing technology; many leaders in the field of molecular biology were in attendance. Although participants concluded that the current technology was insufficient for direct DNA analysis, the conference generated interest in a publicly funded genome project.

In 1985, Charles de Lisi, the new Director of the Department of Energy's Office of Health and Environmental Research (OHER), became interested in the possibility of systematically collecting the sequence of an entire human genome. De Lisi conceived of the idea while reading a 1984 draft report on mutation detection technologies from the Office of Technology Assessment, a technology advisory committee for the U.S. Congress. As the idea was discussed within Department of Energy circles, it was also recognized that such a project could greatly benefit the DOE by improving the public image of, and maintaining a post–Cold War use for, the high-tech nuclear weapons laboratories at Los Alamos. In early 1985, de Lisi convened a workshop in Santa Fe, New Mexico, to garner scientific support, and three months later he began a long process of obtaining congressional funding.

FROM DNA SEQUENCING TO GENETIC MAPPING

At a meeting in Cold Spring Harbor in June 1986, molecular geneticists began to debate the feasibility of a genome sequencing project, spurred by a recent article in the journal *Science*, written by Renato Dulbecco, a co-winner of the 1975 Nobel Prize in physiology or medicine for his work in cancer genetics. Dulbecco argued that a sequencing initiative would benefit cancer research. The article was the first to stimulate discussion within the broader scientific community regarding the desirability and the feasibility of a genome project.

Participants at the 1986 Cold Spring Harbor meeting were divided on the idea; given the apparently massive costs of such a proposal, many scientists were convinced that valuable resources would be pulled away from their own research interests. Many thought of the recently publicized Department of Energy proposal, which almost exclusively emphasized sequencing, as a political tool that was neither useful nor feasible and that would be under centralized control by the DOE. In the process, participants redefined the project and emphasized construction of genetic and physical maps as a more useful initial goal. The conference and its subsequent published proceedings were pivotal in creating scientific support for a genome project and for defining the subsequent process.

The Howard Hughes Medical Institute (HHMI) also played a significant role in the early years of the HGP. The HHMI was founded by the enigmatic American industrialist Howard Hughes Jr. (1905–1976). As one of the largest philanthropic organizations in the world, with assets exceeding $11 billion in 2003, the HHMI has had a long history of supporting genetics research in the United States. Charles Scriver, a Canadian geneticist who was instrumental in initiating a Canadian project, was a member of the Institute's Advisory Board; in 1985, Scriver persuaded the HHMI's Board of Trustees to commit to funding, for the five-year period 1986–1991, the production of genetic databases. Meetings to discuss the Howard Hughes Medical Institute's involvement occurred approximately at the same time as Department of Energy planning meetings, but they had no relation to each other at the outset.

The Howard Hughes Medical Institute played a large part in creating momentum for a genome project. At a July 1986 HHMI-organized conference, Scriver and other leading biologists, in cooperation with representatives from the National Institutes of Health, produced a preliminary plan for a genome project that emphasized genetic and physical mapping, in contrast to the DOE's emphasis on sequencing. It was suggested that sequencing could be done later, when maps were sufficiently detailed and technology had improved. This plan was fine-tuned by many of the same people at a later NIH meeting (discussed later). The HHMI also produced a document in 1987 that described the project, which helped garner congressional support. Additionally, the HHMI also funded White's genome mapping efforts at the University of Utah in the mid-1980s.

Finally, following a meeting, in April 1988, that suggested the creation of an international Human Genome Organization (HUGO), the HHMI funded a September meeting in Switzerland of 32 scientists from 11 countries, at which the participants created HUGO's structure. HUGO was created as a multi-node coordinating program, with one office each in the Americas, London, and Japan. In 1990, HHMI funded the Human Genome Organization with a $1 million startup grant for four years and provided an office and an administrator for the Americas location. The Howard Hughes Medical Institute remained a prominent supporter for genomics in the United States and internationally, playing a pivotal role in the HGP's growth and success.

THE NATIONAL INSTITUTES OF HEALTH

James Wyngaarden, Director from 1982–1989 of the U.S. National Institutes of Health (NIH), had heard about the Department of Energy plan for a project by 1986. In June of that year, he organized a meeting of NIH staff members to discuss a potential role for the NIH in a genomics initiative. The report of the NIH meeting suggested assessing scientific opinion about the project's desirability and announcing NIH's interest at the upcoming (July 1986) conference sponsored by the Howard Hughes Medical Institute. In October 1986, at an NIH Director's Advisory Committee meeting, in which many of the same genome researchers who had been present at the HHMI meeting participated, a proposal was drafted for a National Institutes of Health genomics project that emphasized genetic mapping, the study of model organisms, and improvement in information-handling software. A working group was organized, which began actively creating an NIH program. Wyngaarden approached Congress for funding in early 1987 and received approximately $20 million in 1988. In August 1988, the NIH formed the Program Advisory Committee on the Human Genome in order to advise on the structure of an NIH genome program, and, in October of that year, the NIH formed the Office for Human Genome Research to oversee NIH genome research activity.

By 1987, both the NIH and the DOE were receiving funds from Congress for genomics programs. The existence of two separate genomics initiatives led to arguments that the genome program should be under the control of one administrative structure. The NIH had a long history of involvement in biomedical research, and most scientists preferred its leadership of a public genome project to oversight by the Department of Energy, which some scientists dismissed as "unemployed bomb-makers." In 1987, the National Research Council, a division of the National Academy of Sciences that provided science advice to the federal government, produced a report that argued for the value of a national genomics program that would emphasize physical and genetic mapping and that would consider sequencing only when it became more feasible. The report recommended that there be one lead agency but did not suggest either of the two alternatives as preferred. DOE usually referred to itself as the lead agency, but in 1988 Wyngaarden publicly argued that the NIH should be at the helm. In the spring of 1988, a report of the Congressional Office of Technology Assessment recommended a cooperative project. However, when the OTA recommendation seemed likely to become law, which would force the NIH and the DOE to form a cooperative structure, the two agencies preempted Congress's attempts and created a jointly administered program. They signed an agreement in late 1988, and during the following year an interagency infrastructure was created by participants from the NIH, the DOE, HHMI, and elsewhere.

The resulting NIH-DOE Human Genome Project was announced to the scientific community in 1989 in *Science,* and in a joint report to the U.S. Congress in February 1990. In April 1990, a five-year plan was created for the years 1991–1995, with specific objectives outlined for physical and genetic mapping, bioinformatics (the computer requirements for organizing and analyzing information produced by the project), and the study of model organisms. DNA sequencing was emphasized as a long-term goal, with the assumption that sequencing technology would improve and costs would subsequently decrease.

In the process of discussions about organizing an interagency program, James Watson, the co-discoverer of DNA, became director of the American genome program. The previous year, Watson had become involved in genome efforts at the National Institutes of Health, as head of the NIH's Office of Human Genome Research. He was officially appointed Director of the American Human Genome Project by Wyngaarden in 1988 and remained so until 1992. Following the NIH-DOE agreement, Watson's office became the National Center for Human Genome Research (NCHGR). In 1997, the NCHGR became a full research institute within the National Institutes of Health and was consequently renamed the National Human Genome Research Institute (NHGRI).

The DOE and the NIH continued to maintain separate genome administrative infrastructures using congressional funding. The two programs coordinated their efforts by loosely dividing tasks between them, with some areas of overlap. Congressional funding increased yearly, but, according to the second five-year plan, written in 1993, funding remained below the $200 million per year described in the original five-year plan as a minimum requirement for achieving the goals of the project. The third five-year plan indicated that funding would reach the $200 million mark by 1996.

FUNDING THE STUDY OF ETHICAL, LEGAL, AND SOCIAL ISSUES

Watson stated in 1988 that the HGP would set aside funds for a project to study the ethical, legal, and social issues (ELSI) that might arise from genome research. Approximately 3 to 5 percent of funds for the Genome Project were annually earmarked for ethical, legal, and social issues. An ELSI working group was created within the HGP administrative structure in 1989, chaired by Nancy Wexler, who had studied the psychological effects of the threat of Huntington's disease on individuals and their families. Wexler also had experience with the ethical and social issues surrounding such technologies as genetic diagnostics.

There were two components of ELSI (with a significant amount of overlap): the centralized ELSI infrastructure, consisting of official committees within the NIH and the DOE formed to administer ELSI funding and management projects, and the broader ELSI funding mechanisms for academics who address topics judged as relevant to the ELSI mandate. From its inception, ELSI research has served to examine the potential excesses of claims regarding genomics and to discuss possible abuses of genome research. The work has been broad and diverse, covering such issues as privacy and potential discrimination against those who test positive for genetic diseases or predispositions; the psychosocial effects of genetic testing; the development, marketing, and utilization of genetic diagnostic technologies; commercialization; genetic determinism and other philosophical issues in genetics; and the potential impacts of genetically modified foods.

ELSI is a component of the HGP, and it was created primarily to address potentially negative side effects of the Project. The creation of ELSI deserves credit for being a farsighted and thoughtful investment in studying the broad social impact of the HGP. But it is important to note that the vast majority of ELSI research is not critical of the existence of the project. Instead, there are considerable shared views, goals, and interests between the ELSI initiative and the HGP itself. The ELSI program was certainly not intended to delay or impede the project's progress; instead, it was viewed as an important adjunct that would facilitate its success. Additionally, it is clear from the historical record that the ELSI was seen as a way to avoid the major public controversy experienced during the recombinant DNA debates. In the early years of the HGP, there was a great deal of debate within the biomedical science community about the value of such a large-scale project: many saw it as detracting from small-scale research, as being of dubious, exaggerated value for addressing medical and social issues, and as producing a wealth of potentially dangerous information that society was not yet in a position to deal with ethically. These concerns were frequently expressed in the media, and there was thus an air of controversy surrounding the project. Watson, who has generally been credited with singlehandedly establishing ELSI out of concern for privacy and discrimination issues, in fact realized that Congress was cognizant of these debates between scientists and wanted to see some sort of mechanism for their study and resolution. The commitment of ELSI funding helped allay congressional concerns about the project.

INTERNATIONAL INTEREST IN THE HUMAN GENOME PROJECT

While other countries did have scientists promoting and actively participating in early activities related to the HGP, generally the major influence on the project has been the United States. The United States was the clear leader in genome research, especially with respect to government involvement in creating and funding research and development policies and infrastructures, and its actions spurred other countries to initiate programs of their own. Once these international efforts grew in momentum, however, the influences were multidirectional, as nations fought to establish relative strength in the field and feared being "left behind" in the development of strong genome research programs.

The influence of the American project was often very direct: American institutions often funded international work. The Howard Hughes Medical Institute, for example, provided the majority of funding for the research and operation of the French Centre d'Étude de Polymorphisme Humain (CEPH), an organization created in 1984 at the College de France by Jean Dausset, which standardized genetic mapping by providing almost all genome mappers with DNA clones made from the genomic DNA of a reference set of 40 large multigenerational families. An additional means of American influence involved the active participation of non-American molecular biologists in forming the American program; they subsequently pushed for political commitments to programs in their home countries. As discussed previously, Renato Dulbecco in 1986 wrote an influential editorial arguing that further progress in cancer research required obtaining the complete sequence of the human genome; the article was pivotal in obtaining support from the scientific community for an American initiative. Once the United States had committed to building a genome research infrastructure, Dulbecco organized a sequencing project in his native Italy, of a portion of the X chromosome involved in X-linked mental retardation.

Scientists from several other countries played a similar role: Walter Bodmer, then Director of Research at the British Imperial Cancer Research Fund (1979–1991), participated in early meetings that provided the initial impetus for an American project. Bodmer subsequently prioritized genome research in Britain in his role as co-director of a special committee of the Medical Research Council. Charles Scriver, a Canadian geneticist and a member of the Advisory Board of the Howard Hughes Medical Institute in the United States, was instrumental in attaining federal funding for a Canadian genome program.

But the major influence the United States had on other countries was more indirect; a dawning awareness of growing U.S. economic power in biotechnology, and of the importance to biotechnology of genomics, was a major impetus leading to the creation of genome programs internationally. This direction of influence is illustrated later in this chapter, but first I describe a significant exception to this rule. With regard to Japan, the direction of influence was reversed; one of the main arguments used to spur the American project was the need to remain competitive with the Japanese, and many of the earliest American commitments were a result of a fear of Japanese economic dominance.

JAPANESE GENOMICS

In the early 1980s, Akiyoshi Wada, at the University of Tokyo, began a $4 million large-scale project, funded by the Japanese Science and Technology Agency, titled "Extraction, Analysis and Synthesis of DNA," whose purpose was the development of automated DNA sequencing technology. By 1986, this research, which had by then been moved to the RIKEN Institute, was well known in the United States, prompting discussion of Japan's potential and plans for a genome sequencing project. Wada himself insisted that the technologies were not being designed to facilitate a strictly Japanese project; rather, the primary goal was to produce widely available automated sequencers to encourage rapid DNA sequencing generally.

At about the same time, in 1985, the Japanese Ministry of International Trade and Industry's Agency of Industrial Science and Technology initiated a Human Frontiers Science Program (HFSP), a proposal for international cooperation in the development of human biological research and its key technologies. The project had a modest beginning, with a lukewarm international reception and a relatively conservative commitment by the Japanese government, but by 1990 the project was being seen as a potentially serious large-scale initiative, on a par with the American project.

Some American and Japanese scientists occasionally argued for collaboration between the two countries. Wada had proposed cooperation throughout the period of his research program on automated sequencing technologies, envisioning giant sequencing centers that would serve the international biological community. Wada in fact was pushing for a large-scale international genome project by 1986, with visions of extensive American and Japanese collaboration; his efforts were very influential in encouraging American developments—Charles de Lisi, at the DOE, and other HGP supporters at the NIH and elsewhere met with Wada and were encouraged by the possibilities of collaboration. But politics, and in particular economic tensions between the United States and Japan, intervened and made cooperation rare.

By 1989, the HFSP was doing very well and was having an influence on the American project; American supporters used the Japanese investment in genomics to encourage greater congressional support. The Japanese focus on technology development was especially persuasive, given the existing context of fears about the threat of Japanese dominance in high-technology production. In the context of already existing economic tensions between the two countries, competition rather than collaboration thus came to characterize the relationship between American and Japanese genomics.

GENOMICS IN WESTERN EUROPE

Developments in the United States and Japan influenced European countries to begin allocating funding of their own to genome research, but again the context was generally competition rather than collaboration. This scenario in Western Europe was common across the full spectrum of high-technology ventures: Japanese and American dominance was seen as a threat, and by the early 1980s many commentators warned of the potential inability of Europe to compete with the United States and Japan in the global economy. Many proponents of the genome project, including scientists, policymakers, and industry spokespeople, argued that if their countries did not participate, they would fall behind Japan and the United States and miss out on a crucial opportunity to develop and strengthen national research and development capacity.

The American Congressional Office of Technology Assessment (OTA) recognized that many countries were involved in genome research efforts that might compete with American dominance of the field economically and that this in turn required increased U.S. investment. Its 1988 report, which strongly contributed to congressional interest in building a strong, centrally administered Human Genome Project, argued that, while the United States remained relatively strong in basic genomic research, many other countries were well positioned to commercialize products and technologies related to this research. Commercial interests and concerns were therefore central reasons for government funding of the Human Genome Project.

Between 1987 and 1988, when the American NIH and DOE were still debating the best means for centralizing administration, several European countries took significant steps toward creating genome research programs. The French government committed 8 million francs ($1.4 million US) to be allocated to genome researchers by a committee chaired by the Nobel Prize winner Jean Dausset, the founder and Director of the Centre d'Étude de Polymorphisme Humain (CEPH). Dausset's committee was charged with funding large-scale research projects that focused on mapping and sequencing human chromosomal fragments of particular medical interest. Britain's Medical Research Council had created a special committee, co-chaired by Walter Bodmer, to allocate funds to human genome research, with a focus on large-scale sequencing efforts. The committee had by this time allocated funds to Sydney Brenner, at Cambridge, for his research in genetic mapping and sequencing. Italy had committed approximately $10 million to a large-scale project, headed by Renato Dulbecco, that involved sequencing a portion of the X chromosome involved in X-linked mental retardation. And Germany had begun recognizing, politically if not yet financially, the importance of being competitive in genomics; the federally funded Deutsche Forschungsgemeinshaft (DFG), under the leadership of its vice president, Ernst Winnacker, had begun organizing a conference to promote joint efforts to encourage genome research funding.

There was also a growing interest in Europe in attempting a cooperative European Genome Program to compete with the productivity of Japan and the United States, because it was widely recognized that no one European country could likely compete with either of the two leaders. By July 1988, the European Commission proposed a European human genome program, describing it as a health proposal and titling it "Predictive Medicine: Human Genome Analysis." This proposal was initially rejected by the European Parliament, primarily because of the influence of German Green Party members who were deeply concerned about possible abuses of genetic technologies. For many Germans, "predictive medicine" sounded too similar to the goals of the eugenics movement, a sensitive issue in Germany. By late 1989, a revised proposal was written that made no mention of predictive medicine, and this was accepted by the Parliament by mid-1990. The program was allocated 15 million ECU over three years, with 7 percent of these funds earmarked for projects to study the ethical, legal, and social (ELSI) issues that might arise from genome research.

BALANCING COOPERATION AND COMPETITION

In the early stages of discussion about a Human Genome Project in the United States, American scientists, policymakers, and entrepreneurs debated the desirability of an international genomics project. A common theme was how to balance the interests of cooperation and competition. The 1988 report of the Office of Technology Assessment on the Human Genome Project, for example, listed as one of its three issues of interest "how to strike a balance regarding the impact of genome projects on international scientific cooperation and international economic competition in biotechnology." In many cases, international cooperation seemed at odds with economic competitiveness between nations, but economic arguments were made in favor of international cooperation, as well. Given that the United States was planning to invest significant funds in a large-scale genomics project, concern was expressed about American public funds being used to provide scientific data that industries in other countries might use to develop products. From this perspective, it would be to the benefit of the United States to encourage other countries to invest in genome research to ensure that all countries that might share in the economic benefits also shared the costs.

However, the view of the HGP as an industrial initiative that could provide an economic competitive edge to a country complicated the issue of whether and how international cooperation should proceed. As the OTA report also argued, collaboration was somewhat unpredictable: it could reduce the costs of completing the Project, with each collaborating country contributing modest funds, but it might also reduce the profits that each collaborating country might make from resulting products and technologies. Many American pharmaceutical industry representatives were opposed to collaboration, which they thought might reduce American dominance in the health care market. Competition was thus a very powerful driving force.

From the perspective of the molecular biology community, cooperation was usually seen as desirable. Prominent molecular biologists had significant interests in encouraging international cooperation as a way to avoid redundancy and to encourage the most efficient approach to a genome program. Molecular biologists of course recognized that there were economic interests in their work; many were themselves entrepreneurs, and many others no doubt recognized that selling their research as potentially valuable to industry was the best means for ensuring state funding. But commercialization typically was not the primary motivation of scientists, and many argued that such motivations should not get in the way of international collaboration.

THE HUMAN GENOME ORGANIZATION (HUGO)

The growth of genomics in countries other than the United States and Japan eventually led to international coordination, but not cooperation in a full sense because of the wide, and sometimes reluctant, recognition that the project had a significant economic dimension. Many scientists central to the HGP recognized the difficulties and complexities inherent in the economic value of their work and sought to lessen any negative impact that competition might have on their ability to organize an efficient program. Such concerns for ensuring some form of collaboration in the face of the pressure of economic competition were a motivating factor in the proposed formation of a central international coordinating body for the HGP. As James Watson argued in 1988:

> If they wished, either Western Europe or Japan could by themselves take on this project and it must be assumed that they will initiate their own efforts. So a new international body should soon be formed to ensure that collaboration, not competition, marks the relationship between these efforts in various parts of the world. In a real sense, the exact sequence of the human genome will be a resource that should belong to all mankind. So it is a perfect project for us to pool our talents, as opposed to increasing still further the competitive tensions between the major nations of the world. (Quoted in Office of Technology Assessment, p. 152)

At a meeting in Cold Spring Harbor in April 1988, Sydney Brenner suggested the formation of an international body to coordinate the genome efforts of scientists in different countries. The Howard Hughes Medical Institute (HHMI) funded a September meeting in Switzerland of 32 scientists from 11 countries, who created the Human Genome Organization (HUGO).

THE SEQUENCE

Once a stable infrastructure for the project had been established in the United States, genome mapping and sequencing became systematic, organized, and greatly accelerated. Typically, major research institutes in different countries collaborated to produce the various mapping and sequencing of data. Genetic and physical mapping occurred rapidly, and increasingly detailed maps of the human genome appeared on a yearly basis. By the mid-1990s, very detailed genetic and physical maps of the human genome were published.

But the real goal of the human genome project, which captured the imagination of genome researchers, policymakers, and the media, became the acquisition of the complete sequence of a reference human genome. Originally, this goal, initiated by the interest of the Department of Energy in studying nuclear radiation damage, had been viewed with suspicion. However, once the project ceased to be as controversial as it was in its infancy, the sequence data came to be widely viewed as the ultimate goal of the genome program.

The attainment of this goal was announced in June 2000, and the results were released to the public the following year, by two separate groups: the international public project, coordinated by the Human Genome Organization (HUGO), and a private company called Celera Genomics, stationed in Maryland and founded by J. Craig Venter. Venter had previously participated in the American public project, but he left in 1998 to initiate a private-venture genomics program with Celera. The announcement came in the context of increasingly public disagreements between the two groups with respect to whose research was more accurate and how the data should be used and shared.

Only approximately 85 percent of the genome had in fact been sequenced, and the remainder has proved to be highly resistant to the-then current sequencing technology; additionally, the sequence likely contained many errors. The articles describing the public sequence data, and accompanying articles in the same journal, emphasized that "there is some way to go yet." Additionally, the sequence did not include individual genetic differences; it was a representative collection of DNA sequences only. Also, little had as yet been discovered about the majority of the genes that make up the genome, although it was discovered that there were many fewer genes (approximately 30,000) than the original estimate of 50,000 to 100,000. Only 3 percent of the DNA was in fact found to code for genes, and about two-thirds of the total genes had not yet been identified; of those discovered, the function of most were not understood.

Despite these limitations, and unlike the earlier release of genetic and physical maps, the release of the sequence data catapulted the Human Genome Project into public awareness and captured the imagination of scientists, politicians, and the general public. At the White House press conference announcing the achievement, President Bill Clinton stated that "this is the most wondrous map ever produced by mankind" (quoted in Stevens, p.1). The *New Scientist* magazine referred to the date of the announcement as "the day when humankind learned, in a sense, what it is to be human" (Coghlan and Boyce, p. 4). The announcement also generated considerable enthusiasm in financial circles. Celera's stocks soared following the announcement but then quickly fell when the American and British governments insisted that the human sequence data should be freely available to the public.

CONCLUSION

This chapter describes the diversity of interests, opinions, and activities inherent in the formation of the Human Genome Project. Beginning with early academic work in both public and private centers, at the University of Utah and at Collaborative Research, Inc., respectively, the HGP gradually turned into a state-funded program in the United States; these developments encouraged the growth of state funding internationally. Another notable feature of the HGP was its economic importance. Genomics was consistently tied with biotechnology in lobby efforts and policy documents that discussed the importance of national investment in genomics infrastructure.

As mentioned in the introduction to this chapter, the phrase "Human Genome Project" does not sufficiently reflect the reality that various activities were performed in many locations and that multiple "genome projects" often were initiated independently of one another. But the HGP is most clearly an entity in the political domain. It was discussed as an entity, with specific (albeit changing and controversial) goals and a (more or less) centralized set of infra-structures (the Department of Energy and the National Institutes of Health, and HUGO for international coordination) that created a political head for this complex body of activities.

The goals of the HGP grew as the project grew, but the sequence remained a guiding feature; despite its waxing and waning periods of controversy and excitement, sequence acquisition was a compelling beacon for most of its par-ticipants. By the late 1990s, it was the "holy grail," as some put it: the final step in understanding the essence of ourselves.

With the publication of the sequence in 2000, however, another identity of the HGP came to the fore: that of a stepping stone, an infrastructure, on which would be built a body of work that would culminate in a powerful set of knowledges and applications. This new view describes the many activities that have been under way since 2000 that attempt to use the HGP data to better understand biological structure and function. It also provides a compel-ling justification for continued investment in genomics and these related fields of *postgenomic* biology: proteomics, metabolomics, and more. As these fields grow and expand, so, too, do descriptions of the revolutionary potential of ge-netic research, for understanding many issues of broad scientific, medical, and social importance. The next chapter will explore this potential.

10

GENETICS TODAY

INTRODUCTION

By the new millennium, many of the primary goals of the Human Genome Project—building genetic and physical maps, discovering genes, and sequencing the DNA base pairs of human and other genomes—had been reached. While a great deal was still unknown, interest within the molecular genetic and related scientific communities had moved away from studying genome structure and toward a greater understanding of genome function. The study of biological function was of course far from new, and much had been learned about, for example, protein structure and function, the complexities of cellular functioning, and the similarities and differences between the biology and genetics of humans and other animals. But following the publication of the large-scale map and sequence data, there arose increasingly specific and explicit attempts to use this data to understand the complexities of action and interaction among the full set of genes and proteins that make up an organism.

   This chapter describes some of the major developments in genetics and molecular biology that have arisen in the context of the Human Genome Project. Many of these developments also have benefited from major advances in technology, which have increasingly allowed for the analysis of complex sets of data and complex experimental procedures. A major goal of the human HGP was to identify and understand the function of individual genes, but as the vast amount of data collected during the project became available, researchers became increasingly interested in and aware of the need for a *systems approach* to understanding the interaction of large numbers of genes and proteins in a complex and highly regulated biological environment. Much of contemporary biology today reflects this new approach. In the past ten years, a collection of diverse research programs has arisen, with many different goals and outcomes, that are often referred to collectively as *postgenomics* in order to highlight the goal of studying the function of the entire genome and its products.

POSTGENOMICS

Postgenomics refers to a collection of new research programs that, in a systematic, large-scale, and often comparative way, seek to use data gathered as part of the Human Genome Project in order to understand very complex functional issues. As a whole, postgenomic researchers try to understand the complexity of function and biological meaning of large sections of genomes, or even the entire genome or set of genomes. Traditionally, studies within molecular biology and biochemistry have focused on understanding individual or small groups of molecules. But the enormous amount of data produced by the HGP, combined with technological advances in computing and the resulting advances in managing very complex data, has resulted in many "whole-genome" approaches to traditional studies of genes and their products by understanding the interplay of the vast numbers of molecules operating within complex systems in the cell.

Many other terms have been coined under the general rubric of postgenomics. *Proteomics* and the *proteome* were coined in 1994 by Marc Wilkins, an Australian Ph.D. student looking for a simple way to refer to the *protein complement* of the genome, or the full set of possible proteins for which the genome encodes. Proteomics as a research program is similar to traditional protein chemistry studies, and its goal is similarly to understand protein function. But the approach, like other areas of postgenomics studies, is to understand protein structure and function within a vast and complex interacting system and to use advanced technologies and computer systems to understand these interactions.

While proteomics is the dominant and most encompassing research program within postgenomic studies, other programs have also arisen in the wake of the HGP. Since the term *proteome* was coined, many other postgenomic terms have followed: *transcriptomics* refers to the study of the *transcriptome*, the entire set, or large portions, of possible messenger RNAs resulting from a genome; *metabolomics* refers to the study of the *metabolome*, large sets of metabolic processes under the control of the full set of enzymes coded by the genome. Practically every field of molecular research has seen the rise of an "-omic" term derived from that field's name, in order to emphasize a new systemic and large-scale approach to understanding the traditional topics studied in that field.

"Systems biology" is another new term used to capture this new approach to studying biological function. Systems biology, as described by its advocates, promotes an understanding of cells and organisms as systems and seeks to uses the various "-omic" fields in an interdisciplinary way in order to understand these systems in all of their complexity.

STEM CELL RESEARCH

Perhaps one of the most visible and well-known fields of research within contemporary biology is stem cell research. Stem cells, first discovered in the early twentieth century, form all of the specialized cells of the body. They are *pluripotent*, which means they have the potential to form a variety of specialized cell types. This ability has resulted in a tremendous amount of medical interest in stem cells, since, at least theoretically, they can be used as a source of specialized cells for patients that need them. Skin grafting, organ transplantation, treatment of muscular diseases, the regeneration of nerve tissue, and the treatment of cancer are just some of the medical procedures that could potentially be revolutionized if stem cells could be harnessed for medical use.

The most fundamental form of stem cell is the early embryo. When a sperm and an egg cell fuse to form a zygote, that initial cell is *totipotent:* it can, and does, form all of the cells of the human body. Once this first cell begins to divide a couple of times, the new cells are also totipotent, capable of forming entire embryos if they separate from each other (this is how identical twins are formed). Continued cell division forms an outer layer of cells that will become the embryo's placenta and an inner layer of pluripotent cells (the *inner cell mass*) that will form the various tissues and organs of the embryo.

In 1981, two researchers, Gail Martin, at the University of California, and Martin Evans, at the University of Cambridge, isolated mouse embryonic stem cells. Almost 15 years later, in 1995, James Thompson, of the University of Wisconsin, isolated cells from the inner cell mass of Rhesus monkey embryos. In 1998, Thompson did the same for human embryonic stem cells and managed to grow them indefinitely in a growth medium in the laboratory, thus creating what are now called embryonic stem cell lines.

Thompson obtained his embryonic stem cells from frozen embryos that were left over from private fertility clinics. These clinics use a technique called *in vitro* fertilization in order to help women become pregnant. *In vitro* fertilization involves fertilizing a patient's eggs *in vitro,* outside her body, typically in a small laboratory dish. Typically, more than one egg is fertilized to ensure success, and then a subset is implanted into the patient's uterus. The remainder of the fertilized embryos are either discarded or frozen for potential future use.

Fertility clinics typically have thousands of such frozen embryos that are left over from *in vitro* fertilization procedures, and these are of great interest as a source of embryonic stem cells. Unsurprisingly, however, the research has been a source of major controversy. A great deal of social, religious, and political debate has surrounded the use of discarded embryos for research purposes. In 2001, President George W. Bush banned the production of new embryonic stem cell lines using funds from federal research grants but allowed the use of existing lines for research purposes. American scientists were strongly opposed to the legislation. The issue has been highly divisive politically in the United States; during the 2004 presidential election, for example, the Democratic candidate, John Kerry, argued that Bush's ban was harming American scientific leadership and promised to overturn the ban, and to increase federal science funding more generally, if elected. More recently, a new bill seeks to overturn the ban, and, in 2005, in a move clearly in opposition to federal views, the state of California passed a referendum that allocated $3 billion to stem cell research, to compensate for the lack of federal funding.

The stem cell research community was faced with a major scandal in late 2005 when it became clear that one of the world's leading stem cell research groups, led by Hwang Woo-suk of South Korea, had published fraudulent results of their stem cell experiments. A year earlier, Hwang had reported on the production of human embryonic stem cells tailored to individuals by cloning them from adult human cells. The announcement was considered a major milestone in stem cell research, and it generated enormous excitement in scientific circles and in the media. But within a year it became clear that all of Hwang's research had been faked. The discovery shook the scientific community and has resulted in calls for increased ethical oversight of stem cell research. In response, in 2006, an International Consortium on Stem Cells, Ethics and Law generated guidelines for future stem cell research.

GENOMIC MEDICINE

In the wake of the Human Genome Project, genome research has increasingly been described as an important component of research on a wide range of illnesses, including cardiovascular disease, obesity, diabetes, cancer, mental illness, and infectious disease. The most important public justification for the Human Genome Project has in fact consistently been its potential for broad application to human health improvement, and proponents of the project argue that the results of the HGP will revolutionize medical practice.

Commitments to the power of genome research to revolutionize medicine are typically based on the argument that individuals vary genetically in their susceptibility to many of the leading causes of death and ill health. These diseases are often described as *complex* or *multifactorial:* multiple factors, both genetic and environmental, combine in complex ways to produce a disease state. Individuals with certain gene variants, or alleles, are described as having a *genetic susceptibility:* that is, they are at an increased risk for the given disease because of their genetic makeup. The connection between susceptibility alleles and a given disease is understood as loose and not definitive: most people with susceptibility gene variants will not get the disease, and many people with the disease will not have such variants. The language is that of risk and susceptibility. A study of genetic *risk factors* is argued to be of future use for both preventive and therapeutic disease intervention: individuals with such risk factors, identifiable through the use of genetic diagnostic tests to detect genetic susceptibility, can reduce risky behaviors or possibly obtain preventive therapeutic treatment prior to the onset of disease state, if the increased risk is significantly high.

Traditionally, most genetic testing has been for rare Mendelian conditions. Since the early 1990s, however, several diagnostic technologies have been marketed that detect genes associated with complex diseases. BRCA1 and BRCA2 for hereditary breast and ovarian cancers and HNPCC (hereditary nonpolyposis colorectal cancer) for hereditary colon cancers are several prominent examples. These are familial forms of these cancers, constituting approximately 5 percent of their total incidence; the remaining incidence rates for both forms are a result of nonhereditary, sporadic cases. Individuals with these gene variants have a significantly increased risk of developing the associated conditions, and those without the variants have no risk (or, more accurately, a risk equal to that of the general population for sporadic cancer development).

While an important medical concern, these conditions are not broadly relevant to the general population, although they may in fact have broader relevance if they are marketed inappropriately to those with nonhereditary forms of these cancers, due to pressures on private testing facilities to expand markets. The real promise of genetics, which constitutes its claims of applicability to broader population health issues, is the ability to diagnose and provide intervention possibilities for common complex illnesses that affect the general population, such as cardiovascular diseases, cancer, diabetes, infectious diseases, and mental health conditions. There are currently very few existing applications of diagnostics for genetic susceptibility to common diseases, although they are beginning to appear on the market, and many more are anticipated in the near future.

This variation in susceptibility to common illnesses underlies many of the claims to the potential future benefits of genetics to medicine. Proponents have argued that the study of genetic susceptibility has value in treating diseases because it helps researchers understand and reclassify diseases on the basis of their underlying biological cause. Whereas complex diseases have traditionally been classified according to their physical effects, or *phenotype,* genetics, it is widely believed, will facilitate a new classification and treatment based on their underlying molecular biology, in many cases resulting in the formation of new categories for illnesses that are currently grouped together on the basis of phenotype. One example of such an illness is type II diabetes, which is currently defined as elevated blood sugar, without any specific understanding of the mechanisms responsible. Many have argued that Type II diabetes is probably a set of several different independent diseases that will be elucidated using genetic techniques. The same argument is used for infectious and parasitic diseases such as AIDS, malaria, diarrhea, and respiratory infections, the leading causes of death and ill health in poor, nonindustrialized nations. The assumption is that people vary in their susceptibility to infectious and parasitic diseases and that research into this variation will improve therapeutic options;

additionally, genetics is believed to be capable of clarifying and allowing the reclassification of such diseases.

Another area of genomic medicine that currently is given considerable attention is *personalized medicine,* a view that in the near future drugs will be tailored to the unique genetic constitution of an individual (or small sets of individuals), rather than administered to a whole population of patients. Personalized medicine is the goal of research programs such as *pharmacogenetics* and *pharmacogenomics,* fields of research that attempt to detect genetic variation in both individual responsiveness to drugs and individual risks of side effects. In the wake of the Human Genome Project, proponents argue that such personalized medicine initiatives will become greatly accelerated and accurate and that they will be based on a pharmacogenomic whole-genome analysis of relationships between genetic variations (or *polymorphisms*) and differential drug responses. Many molecular biologists argue that, in the future, personalized medicine approaches will be used widely to more accurately target drugs to specific people and will therefore improve drugs' therapeutic usefulness and reduce the chances of dangerous side effects.

GENETIC MEDICINE POLICY INITIATIVES

Genomics has had some important influence in health policy, and a variety of political institutions have been created to promote genomic medicine. For example, in 1997, the U.S. Centers for Disease Control and Prevention (CDC) formed the Office for Genomics and Disease Prevention (OGDP), a body that outlines policy guidelines for incorporating genomics into both the CDC and states' public health strategies. In 2006 the OGDP changed its name to the National Office of Public Health Genomics (NOPHG).

The NOPHG is run by Muin Khoury, an epidemiologist who was previously Deputy Chief of the CDC's Birth Defects and Genetic Diseases Branch. The conversion of the earlier Birth Defects and Genetic Diseases Branch into the OGDP is a perfect example of the general expansion of medical genetics from rare to common diseases and of the growth of genomic medicine.

In 2000, the OGDP published *Genetics and Public Health in the 21st Century: Using Genetic Information to Improve Health and Prevent Disease,* edited by Khoury; Wylie Burke, Associate Professor of Medicine at the University of Washington, a former member of the OGDP and a member of the U.S. Institute for Public Health Genetics and Chair of the U.S. Department of Medical History and Ethics; and Elizabeth Thomson, the Program Director of the American HGP's Ethical, Legal, and Social Implications (ELSI) Research Program at the National Human Genome Research Institute, National Institutes of Health. The book has been a highly influential guide to how genetics can be and has been incorporated into medicine and public health research and practice. It focuses on how genetics is relevant to, and will be used for, human health improvement and how public health agencies can have a role in this process. Khoury outlines the main argument of the book in Chapter One, which is a framework for the remainder. Chapter One states that the goals of genomic medicine are to expand the concept of "genetic disease" beyond the traditional rare, single-gene diseases that medical genetics studied previously and to include all of the major health issues affecting humanity. The view is generally representative of most proponents' thoughts about the potential revolution that genomic medicine offers. It remains a controversial view, and in the next chapter we will see how other researchers and medical practitioners, especially in the traditional fields of public health, have criticized some of these assumptions.

The book as a whole argues that genetics has an essential role in disease etiology and treatment and in public health strategies, and it outlines a strategy for increasing this role while simultaneously ensuring regulatory oversight. The articles that follow the first chapter primarily offer a sampling of genetic research related to complex disease. There are descriptions of genetic susceptibilities related to taste perception and their possible correlations to obesity, to "unhealthful behaviors" such as smoking and alcohol consumption, to common diseases that affect industrial nations, to infectious diseases that affect primarily nonindustrial nations, and more. As a whole, the book advocates the increased incorporation of genetics into medical education and, in general, the incorporation of genetics, not as a subdiscipline of the field but as a central tool used broadly in public health research and activity. In 2001, the OGDP awarded $300,000 per year for three years to the University of Michigan, the University of North Carolina, and the University of Washington to establish "Centers for Genomics and Public Health." These institutions have in turn been active in public health genomics research, policy formation, and public outreach.

There has also been significant international interest in the genetics of global health. In 2002, the World Health Organization published a report called *Genomics and World Health,* which described the potential for genomics to benefit global health problems. The report itself was inspired by the announcement that the human genome had been sequenced and that the HGP was at its completion. The report focuses primarily on developing countries, and genetic susceptibility research is included as an important priority alongside other subjects of research such as pathogen genome sequencing for drug and vaccine development.

These policy initiatives share a view that genetics is a necessary and beneficial component of understanding how and why people get sick and how to make them better. Much of the language is, again, that of risk: genetic risk factors are seen as powerful causative agents, and an understanding of these will lead to better understanding and treatment of all human disease. The risk-factor approach allows a broad and sweeping view of the role of genetics and various genetic technologies in dealing with disease. Genetic epidemiology is a common framework for conceptualizing the use of genetics to understand and treat complex diseases, especially in such fields as public health genetics and public health genomics; but, as described later, genetic epidemiology is just one of a broader and diverse set of genetic and biotechnological tools and approaches for health improvement.

The CDC publication also stated that public health measures should ensure a balanced approach that also utilizes traditional public health improvement strategies, such as education, environmental improvement, and the alleviation of poverty, but generally it and the other developments described here tend to position genomic medicine as offering the primary means for a revolution in improving the health of the general public. Genomic medicine enthusiasts often point to the inadequacy of environmental approaches and see an essential, typically central, and often revolutionary role for genetics. The many statements and actions to this effect—views of genetics as constituting the source of human illness; arguments that genetics is an important component of all human disease and that it is becoming increasingly unnecessary to distinguish diseases as genetic or not; the many statements describing genetics and the Human Genome Project as precipitating a revolution in health care; and the significant interest and prioritizing of genetic research in various health policy initiatives—all make clear that genetics is broadly viewed not as simply another component in the toolkit of disease treatment and prevention but as a primary, necessary, and revolutionary element in the study, treatment, prevention, and reconceptualization of health and illness.

BIOTECHNOLOGY AND HEALTH

To understand the increasing popularity of genetics in understanding and treating illness, it is important to recognize the economic context in which such views have arisen. That context is the biotechnology industry, introduced in Chapter 8. Since its early origins through companies like Genentech, Amgen, Genzyme, and Biogen, biotechnology has grown to be an important economic sector, and in the United States in particular it has found a valuable role as a generator of new and innovative health products, marketed by the companies themselves or in partnership with multinational pharmaceutical corporations. Health is thus an industry as well as a strictly medical issue, and the economic potential of the biotechnology industry at least partly explains why biotechnology is increasingly used for health improvement.

Many critics have argued that other means of health improvement, such as improving education, cleaning the environment, and decreasing the growing gap between the rich and the poor, are currently being neglected in favor of more biotechnological approaches. It is important to note that biotechnology has produced products that have resulted in real and dramatic improvements in health for many people, but it is also clear that the financial success of the biotechnology industry has helped it to become a popular topic in policy initiatives focused on health improvement.

Various policy papers have discussed how biotechnology might improve health. For example, a report titled "Top Ten Biotechnologies for Improving Health in Developing Countries" was written in 2002 by authors in two programs at the University of Toronto Joint Center for Bioethics (JCB), the Canadian Program on Genomics and Global Health, and the Program of Applied Ethics and Biotechnology. The report's authors convened a panel of 28 biomedical researchers with expertise in both biotechnology and global health issues, who arrived at a consensus about the ten biotechnologies they believe are most relevant to global health:

1. Modified molecular technologies for affordable, simple diagnosis of infectious diseases
2. Recombinant technologies to develop vaccines against infectious diseases
3. Technologies for more efficient drug and vaccine delivery systems
4. Technologies for environmental improvement (sanitation, clean water, bioremediation)
5. Technologies for sequencing pathogen genomes to understand their biology and to identify new antimicrobials
6. Technologies to permit female-controlled protection against sexually transmitted diseases, both with and without contraceptive effect
7. Bioinformatics to identify drug targets and to examine pathogen-host interactions
8. Technologies to promote genetically modified crops with increased nutrients to counter specific deficiencies
9. Recombinant technology to make therapeutic products (e.g., insulin, interferon) more affordable
10. Combinatorial chemistry for drug discovery

Some of these technologies are currently in use (e.g., genetically modified foods, pathogen sequencing, drug development), whereas others are not, and the feasibility of many of these as contributing to global health solutions remains a heavily debated question. Like the genetic risk factor approaches, these approaches have been controversial, and some observers have criticized them as neglecting the more fundamental social causes of poor health in poor countries. But reports such as this illustrate the growing popularity of biotechnological approaches to health care.

EVOLVING CONCEPTIONS OF THE GENE

The considerable advances described in this chapter have had profound effects on contemporary understandings of the gene. The rise of biotechnology and the Human Genome Project have revealed in great detail the molecular structure of DNA, and postgenomic studies have revealed an increasingly complex and interactive system of genetic regulation, with many sorts of interactions among genes and between genes and proteins. Many have argued that this "systems approach" to understanding the gene is fundamentally different from the way the gene was understood for most of the previous century.

Francis Crick once described what he called the "central dogma" of molecular biology: the direction of information and control in molecular biology flows in one linear direction, from DNA to RNA to protein. DNA is transcribed to make RNA, and RNA is translated to make protein. (Crick later claimed that he misunderstood the meaning of the word "dogma," thinking it meant "hypothesis" rather than its actual meaning, "an authoritative principle considered to be absolutely true.") It is increasingly believed that this view of how molecular biology works requires radical revision in the postgenomic era. Given the new understanding of the complexities of the genome and its interactions with the transciptome and the proteome, biologists and philosophers now often contrast modern and evolving understandings of the gene with the classical molecular concept of the gene as described by Crick's central dogma. It is now known, for example, that there are far fewer genes present in the human genome than was previously thought to explain the enormous variety of proteins, biochemical reactions, and other biological functions that occur in the human body. The gene can often produce a variety of different proteins, through a process called "alternative splicing," in which RNA can be modified in various ways after transcription, in order to produce a variety of different proteins. DNA can also frequently be "read" either backwards or forwards by the cellular machinery, producing entirely different proteins in each case. Genes also often encode for only a portion of a protein, so that multiple genes are required for a full protein to be formed.

Beyond this more complex relationship between genes and proteins, it has also become increasingly clear that genes are not the sole source of information and control during the process of protein production. In many circumstances, they are themselves controlled and regulated by proteins and RNA. As discussed in Chapter 7, since the late 1960s it has been known through studies of the lac operon that genes can be regulated by proteins. This process has become increasingly understood, and it is now known that such complex regulatory processes are the norm in cells. What these regulatory processes mean for the concept of the gene is that there is much more involved in the production of a final protein product than simply the so-called gene that supposedly "encodes" the protein. The extent to which our concept of the gene requires modification in the wake of these developments remains a subject of considerable discussion and debate.

CONCLUSION

The sequencing and gene identification initiatives of the Human Genome Project have spurred a focus on understanding how the functions of the proteome, the full set of proteins encoded by the genome, are related to the genome itself. While these sorts of studies were a constant part of scientific activity prior to the HGP, what is new is the focus on large-scale systems and on how large sets of genes, RNA transcripts, and proteins interact with and regulate one another.

The rapid growth of biological data and theory has itself affected contemporary views of the concept of the gene. Today the gene arguably cannot simply be summarized as a unit of hereditary information, as it has been understood since the rediscovery of Gregor Mendel's research in 1900, which was given physical reality by the work of Thomas Hunt Morgan and his colleagues. As the molecular complexity of large-scale DNA-RNA-protein interactions becomes increasingly understood, our understanding of the gene evolves accordingly. The molecular machinery involved in heredity and development is now commonly seen as highly nonlinear and interactive, rather than fitting the simpler model as described in Francis Crick's famous "central dogma."

With the growth of molecular biology has come increasing controversy and debate about its successes and its directions. Major new research programs, such as stem cell research and genetic studies of common illnesses, have been of increasing interest not just scientifically but also socially and politically and have, like the HGP itself, made biology into a very publicly visible scientific activity. In the next chapter we examine the increasingly public nature of molecular biology and some of the currently debated topics and issues that have arisen from modern molecular genetic research.

11

THE GENE AND ETHICAL ISSUES

INTRODUCTION

Since the discovery of the structure of DNA, genetics has become an incredibly successful science, with increasingly complex practices, broad social visibility, major government funding, strong industry interests, and increasing numbers of advocates who argue that genetics will address some of the most fundamental problems and needs of humanity. Genetically modified foods are believed by many to offer a solution to global poverty; genetic studies of diseases are believed to offer a means to provide powerful diagnostic and therapeutic tools for dealing with the major diseases that affect both rich and poor nations; the biotechnology industry is seen by many as the key to generating new drugs and treatments, many tailored to an individual's genetic makeup, that will usher in a new era of personalized medicine; and many believe that genetic studies will shed light on individual differences in all sorts of complex behaviors and character traits, such as intelligence, aggression, food, alcohol and drug cravings, and mental illness.

To quote Spiderman's Uncle Ben, "with great power comes great responsibility." The increasing scientific, social, political, and economic power that genetics has attained has resulted in a great deal of focus on the ethical dimensions of modern genetics. Many research programs have arisen that attempt to understand, anticipate, and alleviate unexpected negative consequences of modern genetic research. Some of these issues are described in this chapter.

ETHICAL ISSUES IN MODERN GENETICS

Genetic Discrimination

With increasing understanding of the genetic basis of diseases, there is concern that people who test positive for some particular genetic defect might suffer discrimination from organizations such as insurance companies and employers. There are many rare genetic diseases that can now be predicted using genetic tests, prior to the onset of any of the disease's physical symptoms. Huntington's disease, for example, causes a progressive degeneration of nervous tissue, resulting in impaired mental abilities, personality changes, depression, and eventually loss of speech, loss of control of motor functions, difficulties breathing and swallowing, severe dementia, and death. The age at which these symptoms arise vary from 30 to 50 or later, and their severity can vary greatly. But the disease is caused by one (mutated) gene, which can be tested for at any point during life (and, indeed, even before birth). Should companies have access to information that an individual has such a mutation? Can an employer decide that such individuals are not an asset to a company because of their future physical state? The issues are complex, and often the rights of individuals and the rights of others might not coincide. For example, many have argued that insurance companies and employers have a legitimate right to know about any genetic conditions individuals might get in the future when considering whether to insure or hire these individuals. Insurance companies are currently allowed to deny insurance to individuals because of factors related to future health. One might therefore argue that they have an equally legitimate right to know about individuals' future risk of genetic disease.

In 2000, the U.S. government prohibited the use of genetic screening of federal employees for purposes such as deciding on what benefits they should get and classifying employees into pay scales; it has also prohibited the disclosure of genetic data about employees to outside agencies. Currently there are no laws against private employers performing any of these actions, although many bills have been proposed in the past five to ten years. None have yet become law, but there are still bills pending that seek to protect people against these sorts of actions. But the issues remains very debated, and the outcome of these bills is hard to predict.

Genetic Privacy and Confidentiality

Many of the issues pertaining to genetic discrimination are also relevant to privacy and confidentiality. There is concern that individuals' genetic data should be their own and that they alone should be able to decide how it is used. But the issues are again complex, and many have argued that in many cases insurance companies and employers have a right to know such information. However, there apparently have been cases in which such a right to know is much harder to defend. Some insurance companies, for example, have been accused of withdrawing coverage of individuals after genetic testing revealed certain gene mutations. Insurance companies have been known to defend such actions by arguing that the genetic condition was a preexisting condition, despite the lack of symptoms, and that therefore the condition was not covered by the patient's insurance. Is this a valid argument? Does having a mutation but no disease symptoms count as having a preexisting condition? Clearly, such cases cry out for ethical study and legislative regulation.

Another important issue with respect to privacy arises when one family member learns that he or she has a certain inherited genetic mutation, since this result increases the likelihood that other family members might also have inherited this mutation. In such a case, does the individual have an ethical obligation to share the new knowledge with family members? Does this information belong solely to the individual patient in such a situation? These issues are much more difficult to solve using regulation, and they typically involve thoughtful case-by-case analysis by trained counselors.

Psychological Effects and Genetic Stigma

Another issue of concern is the impact upon individuals of discovering that they have a genetic disease. For a condition such as Huntington's disease, diagnosis is 100 percent accurate, but age of onset and severity are unclear and there is no clear treatment. Does it do more harm than good to tell people that, at some point in the future, they will get a debilitating and incurable genetic disease? In addition to the potential psychological impact of genetic testing, there are also potential psychological impacts of being labeled as having a certain genetic disease. Being stigmatized as having a genetic defect can potentially cause a feeling of hopelessness and inevitability that is unique to genetic diseases, as people see genetic diseases as more permanent and unavoidable than other diseases.

In addition to individual stigma, group stigma has been of major interest in ethical studies of genetics. Many studies of disability research, for example, have analyzed medical discussion of genetic-related disabilities, such as Down Syndrome, a form of moderate to severe mental retardation found in individuals who have an extra copy of chromosome 21. These studies generally conclude that medical language tends to emphasize the negative and tragic dimensions of such conditions and the goal of preventing the birth of disabled children as the major priority of counseling and screening programs. Such language and views run the risk of diminishing the value of the lives of existing disabled people.

Another major issue of concern is the relationship between genetic studies and race. While genetic studies themselves have illustrated that the concept of race is outdated and not based on clear genetic differences between groups, race has ironically continued to be a major ethical issue in genetics. Some genetic conditions affect some groups of people more than others. For example, sickle-cell anemia, a condition caused by one mutated gene that results in decreased ability of the blood to deliver oxygen to tissues, affects primarily people of African descent. American screening programs in the 1970s targeted many African Americans without their consent or understanding of the nature of these tests, without any protection of privacy, and without any consultation about how best to approach the issue of what to do once people are given the test results. Similar issues have arisen about other screening programs. A major component of ELSI research is to ensure that screening programs are carried out with the participation of and for the benefit of those who are screened.

THE HISTORY OF ELSI

James Watson, co-discoverer of the structure of DNA and the first Director of the Human Genome Project, stated, in 1988, following announcements that the government would fund that project, that the HGP would set aside funds to study the ethical, legal, and social issues (ELSI) that might arise from genome research. Approximately 3.5 percent of funds for the Genome Project were earmarked each year for ELSI. In early 1989, an ELSI working group was created within the HGP administrative structure and chaired by Nancy Wexler, who had studied the psychological effects of the threat of Huntington's disease on individuals and their families. Wexler also had experience with the ethical and social issues surrounding such technologies as genetic diagnostics.

There are two components of ELSI (with a significant amount of overlap). The first is a centralized ELSI infrastructure, consisting of official committees within the National Institutes of Health and the Department of Energy. These ELSI programs were formed in order to administer ELSI funding and management projects and to distribute grant money to academics who wanted to address topics judged as relevant to the overall ELSI mandate and goals.

From its inception, ELSI research has served to examine the potential excesses of claims regarding genomics and to discuss possible abuses of genome research. The work has been broad and diverse, covering such issues as privacy and potential discrimination against those who test positive for genetic diseases or predispositions; the psychosocial effects of genetic testing; the development, marketing, and utilization of genetic diagnostic technologies; commercialization; genetic determinism and other philosophical issues in genetics; and the potential impacts of genetically modified foods. Some of these issues are described in more detail in this chapter.

ELSI programs were not designed in opposition to the Human Genome Project. ELSI is a component of the HGP, created primarily to address potentially negative side effects of the Project. The creation of ELSI deserves credit for being a farsighted and thoughtful investment in studying the broad social impact of the HGP, but it is not critical of or opposed to the goals and work of the project itself. The ELSI initiative and the HGP itself share many views, goals, and interests, and ELSI was viewed as an important adjunct that would facilitate the HGP's success. As the ELSI Web site states, "The establishment of the ELSI program was viewed as vital to the success of the HGP."

Additionally, it is clear from the historical record that establishing ELSI was seen as a way to avoid the major public controversy experienced during the recombinant DNA debates. As described in Chapter 8, the advent of genetic engineering elicited considerable protest within scientific and public circles, and many scientists were surprised by this response. Watson and others learned a valuable lesson from the recombinant DNA controversies about demonstrating to the public that scientists were taking ethical issues seriously.

In the early years of the HGP, there was a great deal of debate within the biomedical science community (on a scale rarely seen in contemporary discussions) about the value of such a large-scale project. Many saw it as detracting from small-scale research, as being of dubious, exaggerated value for addressing medical and social issues, or as producing a wealth of potentially dangerous information that society was not yet in a position to deal with ethically. Congress was cognizant of these debates between scientists and wanted to see some sort of mechanism for their study and resolution. Watson, who has generally been credited with singlehandedly establishing ELSI out of concern for privacy and discrimination issues, also understood the political necessity of addressing and allaying congressional concerns.

ELSI work has therefore occasionally been criticized by some as being compromised by its funding source, as being unable to objectively assess the purported benefits of the HGP, and instead as functioning as a public relations branch of the HGP, helping to legitimate the large-scale genomic research approach to addressing human medical and social needs. But ELSI certainly isn't simply a propaganda tool: some ELSI-funded work has been highly critical of genomics. Many researchers who have studied the HGP with ELSI funding have been quite willing to describe the program as a complete mistake, causing more social harm than good. But it is still true that the central ELSI infrastructure accepts the basic premise that the HGP will provide enormous benefits to society, without questioning either its underlying utility or purpose or the degree of priority and funding given to a genetic approach to what are actually social and political issues, including poverty, crime, and many public health equalities.

BEYOND ELSI: CRITIQUES OF GENETICS

ELSI research originated as part of the Human Genome Project itself, but other research on the social and ethical dimensions of genetics originated outside the HGP. We have already pointed to some controversial areas of genetic research in prior chapters, such as stem cell research, cloning, and the use of genetically modified foods in agriculture. Each of these has seen critical analysis by commentators who often are not part of the traditional "ELSI establishment" but instead are outsiders to the world of the Human Genome Project and genome research more generally.

From the perspective of this book and our interest in the concept of the gene, one of the most interesting contemporary debates surrounds the degree to which genome research will bring medical advances. In the previous chapter we discussed the rise of genetic medicine, an approach that sees genetics, molecular biology, and biotechnology as offering a means to revolutionize the study and treatment of the most common diseases that affect humanity. But accompanying this trend has been a growing body of literature criticizing genetic medicine as ignoring the broader, social context of most of the world's health problems. These studies are typically much more critical of the fundamental claims of genetic medicine than those studies that fall under the ELSI umbrella. These studies have generally argued that harnessing the HGP data is not an especially effective approach to improving the health of the general population, much less a revolution in medicine.

Several studies have argued that, claims to broad applicability notwithstanding, the gains made from genetics are modest, with relevance to relatively few people with rare genetic diseases, and that even for these rarer conditions, genetics has so far had little to offer in terms of ameliorating suffering and preventing death. Others have argued that, even if claims about the applicability of genetics to treating common illnesses come to fruition, the focus on individuals is an inappropriate route to affecting population health levels. It is commonly argued by critics of genetic medicine that, while it is true that multifactorial diseases involve the impact of a multitude of environmental factors on our individual genetic endowments, in reality the most significant risk factors for common diseases are the environmental factors, not the individual genetic endowments. Various evidence has been used to demonstrate the dominance of the environment. For example, the incidences of common diseases vary enormously in different areas of the world, and migration studies have shown that individuals who move from a low-incidence to a high-incidence area acquire the new, higher risk for that area. The conclusion of these studies is typically that, rather than identifying individuals with an elevated risk due to genetic susceptibility to conditions like diabetes or heart disease, it would be more useful in promoting public health to institute social and environmental changes that reduce the risk for the entire population. Such arguments regarding individual and population approaches to health are commonly based on the work of Sir Geoffrey Rose, an epidemiologist with the London School of Hygiene and Tropical Medicine. In a series of articles published in the mid-1980s and summarized and expanded in his 1992 *The Strategy of Preventive Medicine,* Rose argued that, in order to have an appreciable effect on public health, one must focus on the *cause of incidence,* that is, population-level risk factors, rather than the *cause of cases,* individual risk factors such as suscepti-bility genes. A focus on the causes of cases, Rose argued, can explain disease distribution *within* a population, but it typically misses those factors that act on the population as a whole. As Rose often argued, the hardest risk factor to find is that shared by a whole population.

To illustrate his distinction between causes of incidence and causes of cases, Rose discussed chronic heart disease (CHD), which is strongly correlated with high cholesterol. He described studies of two populations of middle-aged men, in Finland and in Japan, that reveal that, within each population, there are individual differences in serum cholesterol levels, distributed normally in the given population. A small number of individuals within this normal distribution are at a much greater risk than the general population for CHD, and a cause-of-cases approach would typically focus on these individuals as targets for research and treatment. But Rose demonstrates that this approach would miss the fact that CHD is very common for the group of Finnish men, while for the Japanese men it is quite rare, because the serum cholesterol distribution curve in the Finnish group is markedly higher than that for the Japanese group. Rose thus argues that a rational and effective public health approach designed to reduce CHD incidence should target shared causes of high cholesterol levels for the high-risk population, rather than causes for individual, high-risk cases within this group.

Rose sees genetic factors as a dominant example of his causes of cases—that is, as an example of a risk factor that might explain some disease distribution among individuals within a population but that is less important than environmental factors as a cause of risk for populations as a whole. Since Rose's work, many others have critically analyzed genetic medicine from this perspective.

One major environmental effect emphasized in studies of illness is pov-erty, or socioeconomic status (SES), the term normally used. The link between wealth and health has long been a topic of interest and debate. Numerous studies from the early nineteenth century to the present have shown a clear, consistent correlation between health and SES. Several studies are pivotal and have become the foundation for much of the contemporary work in this field. One is the so-called Black Report, written in 1980 by a British working group chaired by Sir Douglas Black. The report concluded that a concerted govern-ment effort to reduce income inequality in society would result in dramatic health improvements for a large percentage of the population.

The Black Report became a landmark study, with broad influence inside and outside public health circles. The Black Report was not the first study to emphasize the correlation between health and wealth, but it was one of the most comprehensive, and it set a standard for subsequent studies. By the 1990s, many researchers had addressed and refined the relationship between SES and health. Since then, it has become increasingly understood that the link between SES and health is not a simple dichotomy of health inequality; instead, there is a gradient, with a consistent effect across entire populations, showing that incremental increases in SES are correlated with incremental decreases in illness and early death. The most commonly cited studies that suggest this trend are the Whitehall Studies, which showed that the health and longevity of a group of middle-class British civil servants showed incremental improvements the higher their position in the civil service hierarchy. Income inequality, rather than absolute wealth, thus appears to be the chief determinant of population health inequalities.

The specific factors behind the link between wealth and health are less clear; a variety of recent studies have argued that individual or lifestyle factors such as exercise patterns, smoking, alcohol consumption, and diet account for some but not the majority of this correlation. Most studies suggest that the link is a result of complex factors related to social and environmental factors such as family, community, and work environments, but the full nature of these intermediate factors remains unclear. Some epidemiologists have questioned the value of searching for intermediate factors, given that the *primary* factor, unequal socioeconomic status, is already known and should therefore be the focus of policy efforts. The Black Report itself called for state measures to reduce income disparities between members of the general population, but the social welfare programs required to achieve this were deemed by Margaret Thatcher's government to be too costly and to represent inappropriate government interference in the British economy.

Some critics have argued that a focus on individual cases, such as by identifying genetic risk factors, is potentially harmful to population health, because it might divert attention and scarce health care funding from the more central environmental and socioeconomic determinants of health. It has also been argued that the commercial value of biotechnology has enabled genetic medicine to grow in popularity within circles of political power, giving its proponents undue influence over how best to approach health issues. As we have seen, biotechnology is a significant area of industrial growth, and genomics is a major component of biotechnology. Political investment in genetic approaches to health is partly a consequence of this commercial interest.

In fact, investment in genetic research is often seen as both a health investment *and* an industrial investment. Modern governments typically work under the assumption that the most efficient way to improve society is to encourage the market to produce products for individual consumers. In contrast, critics of genetic medicine argue that governments need to take responsibility for removing inequalities within populations. The debate over how best to encourage the growth of healthy populations is therefore part of a larger political debate about the role of governments and private industries in social improvement. Those who argue for a focus on *populations* typically suggest that governments need to intervene to reduce income inequalities, using various social welfare measures. In contrast, those who argue for a focus on *individuals* typically advocate, implicitly or explicitly, for market mechanisms that will encourage the growth of health-based industries and the production of health-based products for individual patients as consumers. For example, in the "Foreword" to the World Health Organization's 2002 report titled *Genomics and World Health,* discussed in the previous chapter, Gro Harlem Brundtland, the Director General of WHO, argues that health research in rich nations is market-driven, not in order to suggest avoiding health improvement within such a framework but rather to encourage the extension of the framework to developing countries. The issue of concern is that market-driven health research is not occurring in developing nations and should be encouraged.

Population-level and individualized approaches to health care have different potentials and different limitations. The population approach is clearly key to reaching the broader goals of improving health, but its benefit is diffused throughout society and offers little support for individual disease sufferers. Thus, there is clearly a role for individualized approaches. Despite its inability to address broad population health disparity, the market produces health products that address important individual health needs, albeit often imperfectly. What is needed is political commitment to ensuring that both individual and population-level interventions produce tangible social benefits—a large topic that is beyond the scope of this book.

These critical analyses of the role of genetics in health improvement provide yet another example of how the concept of the gene continues to be a subject of debate. Part of the attraction of genetic medicine is its depiction of the gene as the fundamental cause of illnesses. From the simple models of genetic diseases caused by a mutation in one gene to the much more complex depiction of common illnesses as resulting from complex interaction between specific gene forms and between genes and the environment, the central belief of genetic medicine is that illness can be traced back to the genes and that the genes are the site at which we should focus our efforts to revolutionize medicine. This view in turn rests on a concept of the gene as being the primary cause of all subsequent biological effects, including growth, development, proper biological functioning (including a vast number of physical and mental characteristics), and disease, in cases where this functioning goes awry. This belief stems from Francis Crick's "central dogma" that information flows in one direction, from DNA to RNA to proteins. According to the central dogma, genes cause higher-level biological functioning; therefore, in the case of medicine, (mutated) genes are the fundamental cause of illness. But, as we saw in the preceding chapter, the concept of the gene has been changing dramatically in the past several decades, and genes are increasingly understood as complex components of an interactive network of living processes. The assumption that genes "control" higher functions is increasingly seen as too simplistic. Sometimes proteins and RNA affect DNA, and genetic information is not quite so rigid and controlling as many assume it to be. This new concept of the gene is still a work in progress, but it will likely affect understandings and applications of genetic medicine in the future.

CONCLUSION

With the rapid rise in prominence of genetics since the discovery, in 1953, of the structure of the double helix and the development, in the 1970s, of recombinant DNA and monoclonal antibody technologies, the field of genetics has become the subject of a great deal of public scrutiny. As the gene becomes better understood and is more easily used and manipulated for a variety of purposes that promise medical, agricultural, and economic benefits to humanity, ethicists, lawyers, social scientists, and historians have all become interested in genetic research and its broad and diverse effects. Many of the scientists who have been involved in this sort of research have themselves initiated this sort of social scrutiny. In particular, the initiators of the Human Genome Project in the United States were themselves responsible for creating an Ethical, Legal, and Social Issues (ELSI) program, funded using a percentage of the money allocated to the Human Genome Project by the U.S. government. The ELSI program has resulted in increased understanding and awareness of the potentially negative side effects of the Human Genome Project and has encouraged initiatives to safeguard against potential problems such as genetic discrimination, the loss of privacy, and genetic stigma, while simultaneously supporting the central goals and aspirations of genomic research.

Other social scrutiny has challenged the fundamental tenets of the value of genomics itself. Genetic research has faced opposition from those who view the social benefits of genetic research as being overblown and see its economic dimensions as undermining its credibility. This is particularly evident in fields related to genetic medicine, where critics see genetic research as deflecting attention from more fundamental, population-level approaches to addressing health problems by reducing social inequalities. What is needed is political commitment to balancing individual and population-level health initiatives, while ensuring that specific health research projects truly provide health benefits for individuals and for society. What is also needed is the continued incorporation into genetic medicine of a recognition of the multiple levels of causation and complexity contained at the levels of the gene, the organism, and the environment. In the future, this recognition will likely be a consequence of our increasingly sophisticated understandings of the concept of the gene.

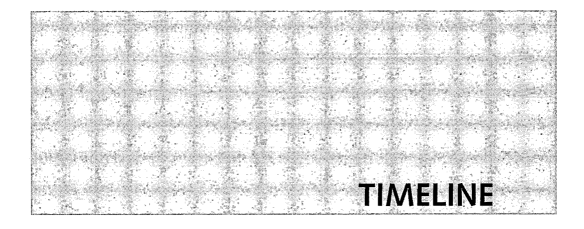

TIMELINE

ca. 355 B.C.	Aristotle founds his school, the Lyceum, in Athens.
170 A.D.	Galen of Pergamum publishes *On the Natural Faculties*.
ca. 180	Galen publishes *On Seed*.
Twelfth to thirteenth century	Ancient Greek texts are translated to Latin and made available to the West.
	The first universities are founded, in Italy (Padua and Bologna), France (Paris), and England (Oxford and Cambridge).
1265	Thomas Aquinas begins writing *Summa Theologiae* (Theological Summary), his most important collection of writings. He continued to work on these until his death in 1274.
1436	Johannes Gutenberg invents the printing press.
1620	Francis Bacon writes *Instauratio Magna* (The Great Instauration).
1628	William Harvey writes *De Motu Cordis* (On the Motion of Blood).
1637	René Descartes writes *Discours de la Méthode* (Discourse on Method).
1651	William Harvey writes *De Generatione* (On Generation).
1664	Robert Hooke publishes *Micrographia*, popularizing microscopic studies.
1672	John Swammerdam describes a preformation theory called *emboitement* in his *Miraculum Naturae* (The Miracle of Nature).

1674	Nicholas Malebranche popularizes preformation theory in his *Search for Truth*.
1685	Antoni von Leeuwenhoek describes, in a letter to the Royal Society, his discovery of *animalcules* in male seminal fluid.
1687	Isaac Newton publishes *Philosophiae Naturalis Principia Mathematica* (Mathematical Principles of Natural Philosophy).
1691	John Ray writes *The Wisdom of God Manifested in the Works of the Creation.*
1740	Charles Bonnett discovers parthenogenesis.
1741	Abraham Tremblay cuts polyps (hydra) in half, and each half regenerates a new organism.
1749	Georges-Louis LeClerc, Comte de Buffon publishes *Histoire Naturelle* (Natural History).
1759	Caspar Friedrich Wolff publishes *Theory of Generation*, describing neural tube formation in a chick.
ca. late eighteenth century	*Naturphilosophie* (philosophy of nature) becomes influential in Germany.
1779	Georges-Louis LeClerc, Comte de Buffon publishes *Epoques de la Nature* (Epochs of Nature).
1794	Erasmus Darwin publishes *Zoonomia.*
1802	William Paley publishes *Natural Theology.*
1809	Jean-Baptiste de Lamarck publishes *Philosophie Zoologique* (Zoological Philosophy).
1817	Georges Cuvier publishes his most important work, *Regne animal distribué d'après son organisation* (translated into English as *The Animal Kingdom*).
1831	Robert Brown discovers the nucleus.
1838/1839	Jakob Mathias Schleiden publishes *Contributions to Phytogenesis* (1838), and Theodor Schwann publishes *Microscopic Investigations on the Accordance in the Structure and Growth of Plants and Animals* (1839), which, together, describe Cell Theory.
1858	Rudolf Virchow writes *Die Cellularpathologie* (Cellular Pathology) and coins the phrase *omnis cellula e cullula* (all cells from cells).
1859	Charles Darwin writes *The Origin of Species.*
1861	Max Schultze identifies protoplasm (now called cytoplasm) in cells.
1865	Gregor Mendel publishes "Versuche über Pflanzen-Hybriden" (Experiments on Plant Hybrids).

1868	Thomas Henry Huxley publishes "On the Physical Basis of Life," popularizing protoplasm theory.
	Friedrich Miescher discovers deoxyribonucleic acid (DNA).
1871	Charles Darwin publishes "Pangenesis."
1883	August Weissman publishes *Uber die Vererbung* (On Heredity), which describes a theory of hard inheritance and identifies heredity and development as independent biological processes.
	Francis Galton coins the term *eugenics* and founds the eugenics movement.
1887–1888	The Germans Theodor Boveri and August Weismann confirm the process of reduction division (meiosis) during gamete formation, in which each gamete receives half of the nuclear material.
	Heinrich Waldeyer coins the term *chromosome* to describe the bundles of nuclear material seen during cell division.
1889	Hugo de Vries publishes *Intracellular Pangenesis*.
1892	Wilhelm Roux publishes *Die Entwickelungsmechanik der Organismen* (The Developmental Mechanics of Organisms).
	August Weismann publishes *Das Keimplasm (The Germplasm)*, distinguishing germ cells from somatic cells.
1894	William Bateson publishes *Materials for the Study of Variation,* advocating discontinuous inheritance.
1900	Gregor Mendel's paper is discovered by Carl Correns, Hugo de Vries, and Erich Tschermak, who recognize its support for a particulate theory of inheritance. Mendel is posthumously considered the founder of this new theory, called *Mendelism.*
1901	Hugo de Vries publishes *Die Mutationstheorie* (The Mutation Theory).
1902	William Bateson publishes *Mendel's Principles of Heredity: A Defence.*
1904	Thomas Hunt Morgan begins a position at Columbia University and forms the *fly lab*, using *Drosophila melanogaster,* the common fruit fly, as a model organism for studying biological processes.
1905	William Bateson coins the term *genetics.*
1909	Wilhelm Johannsen coins the terms *gene* and *genotype.*
	Frans Alfons Janssens (F. A.) Janssens discovers the phenomenon of genetic crossover during cell replication.
1910	Charles Davenport founds the Eugenics Record Office (ERO).

	Thomas Hunt Morgan discovers a genetic linkage between eye color and sex determination in *Drosophila*.
1913	Alfred Sturtevant, working in Thomas Hunt Morgan's fly lab, produces the first genetic linkage map.
1915	Thomas Hunt Morgan publishes *The Mechanism of Mendelian Heredity*.
1917	Karl Ereky coins the term *biotechnology* to describe the use of organisms for industrial production.
1924	Hans Spemann discovers embryonic induction.
1927	In *Buck v. Bell*, the United States Supreme Court upholds the constitutionality of compulsory sterilization of the "feebleminded," a decision that results in the sterilization of Carrie Buck.
1928	Thomas Hunt Morgan moves to the California Institute of Technology and founds its Division of Biology.
1930	Ronald A. Fisher publishes *The Genetical Theory of Natural Selection*.
	Sewall Wright describes the phenomenon of genetic drift.
1931	The Rockefeller Foundation initiates its Science of Man program.
1932	J.B.S. Haldane publishes *The Causes of Evolution*.
1937	Theodosius Dobzhansky publishes *Genetics and the Origin of Species*.
	Julia Bell and J.B.S. Haldane detect the first genetic linkage in humans, between hemophilia and color-blindness.
1938	Warren Weaver coins the term *molecular biology*.
1939	Linus Pauling publishes *The Nature of the Chemical Bond*.
1941	Max Delbrück and Salvador Luria found the phage group
1942	Julian Huxley publishes *Evolution: The Modern Synthesis*.
	Ernst Mayr publishes *Systematics and the Origin of Species*.
1944	George Gaylord Simpson publishes *Tempo and Mode in Evolution*.
	Erwin Schrödinger publishes *What Is Life?*
	Oswald Avery publishes research suggesting that DNA might be the hereditary material in the gene.
1946	The National Institute (now Institutes) of Health becomes a science funding governmental agency and begins a period of rapid expansion and influence
1948	Edwin Chargaff discovers that DNA has an equal number of adenine and thymine bases and an equal number of cytosine and guanine bases.
1949	Sven Furberg determines the three-dimensional structure of a single DNA nucleotide.

1952	Alexander Todd discovers that the phosphates molecules of DNA are linked into a chain.
	John Griffith discovers that the DNA base adenine attracts thymine, and guanine attracts cytosine.
	Alfred Hershey and Martha Chase confirm Avery's suggestion that DNA is the hereditary molecule.
	Rosalind Franklin produces "Photo 51."
1953	Linus Pauling publishes a model for the structure of DNA, which turns out to be incorrect.
	James Watson and Francis Crick publish their celebrated model of the structure of DNA.
	George Palade discovers microsomes (now called ribosomes).
1954	George Gamow founds the RNA Tie Club.
	Francis Crick suggests that a triplet of DNA bases most likely encodes an amino acid.
1955	Arthur Kornberg discovers DNA polymerase.
	Oliver Smithies invents gel electrophoresis.
1957	Matthew Meselson and Franklin Stahl discover semi-conservative replication.
1958	Sydney Brenner, Francois Jacob, and Jacques Monod describe messenger RNA.
1961	Marshall Nirenberg cracks the genetic code.
	Francois Jacob and Jacques Monod discover the lac operon.
1962	Transfer RNA is characterized.
1966	The complete genetic code is elucidated.
1967	Mary Weiss and Howard Green invent somatic cell hybridization.
1970	Herbert Boyer discovers restriction enzymes.
1971	Paul Berg and Peter Lobban independently combine DNA from two different organisms.
	President Richard Nixon declares "War on Cancer," promising a cure by the American bicentennial in 1976.
	Torbjorn Caspersson utilizes chromosome staining.
1973	Stanley Cohen and Herbert Boyer invent recombinant DNA technology, creating the first genetically engineered organism.
	Cohen and Boyer insert the first mammalian DNA into bacteria, using recombinant DNA technology.
1975	Paul Berg convenes the International Congress on Recombinant DNA Molecules, at Asilomar, California (the Asilomar Conference).
	Ed Southern invents the Southern Blot.

1976	The National Institutes of Health Recombinant DNA Advisory Committee releases its guidelines for research using recombinant DNA technology.
	Herbert Boyer and Robert Swanson found Genentech to commercialize recombinant DNA technology.
1977	Boyer, at Genentech, and other colleagues produce the first human protein, somatostatin, using recombinant DNA technology.
	Allan Maxam and Walter Gilbert, and Fred Sanger independently, invent DNA sequencing techniques.
1978	Genentech scientists produce human insulin, the first commercial product made using recombinant DNA technology.
	David Botstein, at MIT, and others discover restriction fragment length polymorphisms (RFLPs).
	Yuet Wai Kan discovers and isolates a RFLP marker that is located 13 cM from the gene for sickle-cell anemia.
1980	Ananda Chakrabarty produces a bacterium engineered to consume oil, the first patent granted for a living organism other than plants.
	The U.S. Congress passes the Patent and Trademark Amendment (Bayh-Dole) Act.
	David Botstein and Robert Davis write a paper advocating the creation of a dense genetic map of the entire human genome.
1981	Gail Martin and Martin Evans isolate mouse embryonic stem cells.
1982	Monsanto Corporation and, independently, several groups of scientists invent a method to insert genes in plants, which is subsequently used by Monsanto to produce the first genetically modified foods.
1983	Philip Leder of Harvard University invents the Oncomouse™, a mouse genetically modified to be highly susceptible to cancer.
	Akiyoshi Wada receives $4 million from the Japanese Science and Technology Agency to develop automated DNA sequencing technology.
1984	John Sanford invents the gene gun (biolistics).
	Jean Dausset founds the Centre d'Étude de Polymorphisme Humain (Center for the Study of Human Polymorphisms), or CEPH.
1985	Kary Mullis invents the polymerase chain reaction (PCR).
	Charles De Lisi, at the Department of Energy, organizes a conference to explore methods for large-scale sequencing of the human genome.

The Howard Hughes Medical Institute commits to five years of funding the construction of databases for storing DNA sequence information.

The Japanese Ministry of International Trade and Industry's Agency of Industrial Science and Technology initiates the Human Frontiers Science Program (HFSP).

1986   Leroy Hood invents the automated DNA sequencer.

Several meetings are held by the National Institutes of Health and the Howard Hughes Medical Institute to outline a proposal for a human genome project.

Renato Dulbecco writes an editorial in *Science* advocating a genome project.

1987   Maynard Olson and David Burke invent yeast artificial chromosomes (YACs).

The British and French governments create committees, chaired by Walter Bodmer and Jean Dausset, respectively, to fund genome research.

1988   The Human Genome organization is founded by Sydney Brenner, Victor McKusick, Leroy Hood, Walter Gilbert, and other leading molecular biologists.

The National Institutes of Health forms the Office for Human Genome Research.

The National Institutes of Health and the Department of Energy merge their separate genome research activities, officially launching the American Human Genome Project.

1989   An ethical, legal, and social issues (ELSI) working group is formed as part of the American Human Genome Project.

Maynard Olson, Leroy Hood, David Botstein, and Charles Cantor develop sequence-tagged sites.

The European Commission creates a European Union genome program.

1994   Monsanto Corporation markets bovine growth hormone for increasing milk production in dairy cows.

Calgene markets the Flavr Savr™ tomato.

Marc Wilkins coins the terms *proteome* and *proteomics*.

1995   Monsanto markets *Bt* Corn, genetically modified to be resistant to multiple insects.

1996   Monsanto markets soybeans genetically modified to be resistant to Glyphosate, a Monsanto herbicide.

1997   The U.S. Centers for Disease Control and Prevention forms the Office for Genomics and Disease Prevention (OGDP).

1998   James Thompson creates human embryonic stem cell lines.

2000	The American Human Genome project and Celera Genomics announce the completion of a draft of the human genome sequence.
	The U.S. government prohibits genetic screening of federal employees.
2001	George W. Bush bans the production of new embryonic stem cell lines using funds from federal research grants.
2002	The World Health Organization publishes *Genomics and World Health*.
	Bioethicists at the University of Toronto Joint Center for Bioethics publish "Top Ten Biotechnologies for Improving Health in Developing Countries."
2004	Hwang Woo-suk publishes research claiming that his laboratory successfully created human stem cells from adult tissues.
2005	The state of California passes a referendum that allocates $3 billion to stem cell research.
	Hwang Woo-suk's research is discovered to be fraudulent.
2006	An International Consortium on Stem Cells, Ethics, and Law generates guidelines for future stem cell research.

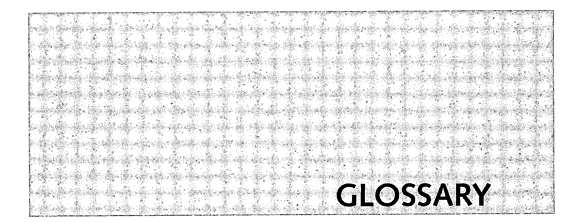

GLOSSARY

additive genetics: a term describing multiple genes that, together but with no interaction, form a physical trait.

Age of Love: a theoretical time period during which the forces of attraction were dominant in the universe and all living things were formed mechanistically; advocated by Empedocles in the fourth century B.C.

amino acid: a subunit of proteins.

animalcules: Antoni van Leeuwenhoeks's term for sperm cells that he observed under his microscope in 1685. Leeuwenhoek, a preformationist, assumed that these were fully formed organisms. See: preformationism.

apomictic: a term describing a higher organism that can reproduce asexually as well as sexually.

argument from design: an argument stating that the world and its contents are much too complex to have come into existence by accident or randomly; therefore, there must exist a being outside the world that created such complexity. See: natural theology, divine watchmaker.

atomism: an ancient Greek theory of mechanism that stated that all of reality is composed of, and can be explained by, units of matter that are in constant, random motion. See: mechanism.

bacterial artificial chromosome (BAC): a laboratory-constructed chromosome that can be inserted into bacteria. Used for inserting large amounts of foreign DNA into bacterial cells. See: chromosome.

biochemistry: the study of chemical reactions inside the cells of organisms.

biolistics: see: gene gun.

biology: the study of living things. The term is usually attributed to Jean-Baptiste de Lamarck, who popularized it.

biometrics: the statistical analysis of populations of organisms.

biotechnology: the use of organisms for industrial production. Often used to refer to processes based on recombinant DNA technology. See: recombinant DNA technology.

blastula: an early-stage embryo that has yet to begin differentiating various tissues and organs.

blending theory of inheritance: the theory that inheritance involves the mixing together of minute particles of matter from both parents, which results in an offspring with a form intermediate to that of each parent.

cause: in Aristotelian philosophy, a component of a proper explanation of reality. There are four Aristotelian causes: the *material cause,* or the matter from which an object is formed; the *efficient cause,* or the creator of an object; the *formal cause,* or the form of an object; and the *final cause,* or the purpose of an object.

cause of cases: individual risk factors for a given illness.

cause of incidence: population-level risk factors for a given illness.

cell theory: the theory described by Jakob Mathias Schleiden and Theodor Schwann, stating that cells are the fundamental units of living organisms and that all cells come from preexisting cells. See: nucleus, chromosome.

chromosome: structure in the nucleus of the cell, consisting of proteins and deoxyribonucleic acid, and shown by Thomas Hunt Morgan to be the site of the genes. See: nucleus.

colloid: now obsolete, but once believed to be an aggregate of small molecules held together by an unknown vital substance unique to living things.

complex trait: a physical trait describes as a product of complex interactions between multiple genes and an organism's environment.

continuity of the germplasm: August Weismann's argument, made in the late nineteenth century, that there exist germ cells that are distinguishable from somatic cells, that are responsible for hereditary transmission, and that are unaffected by any environmental effects upon the rest of the body. The theory was the first to separate heredity from development. See: germplasm.

continuous evolution: the theory that continuous variation in a population is the source of evolutionary change. See: evolution, Darwinism, Mendelism, discontinuous evolution.

correlation of parts: Georges Cuvier's theory stating that an organism is perfectly designed to function as a whole in its environment. Any change in one part of an organism would immediately result in drastically reduced fitness. Cuvier thus denied that evolution was possible: organisms were too well adapted to their environments to allow room for changes in their structure.

Darwinism: Darwin's theory of evolution by natural selection of organisms from a population exhibiting continuous variation. See: continuous evolution, discontinuous evolution, evolution, Mendelism.

demiurge: a being, conceived by Plato, that created the material world as an imperfect replica of the true, immaterial world of ideas.

deoxyribonucleic acid (DNA): one of the two components of chromosomes (the other is protein); discovered to be the molecule that encodes the gene. See: gene, chromosome.

discontinuous evolution: the theory that evolution occurs by way of sudden changes or obvious differences in variation within a population. Advocated by Mendelists as an alternative to the Darwinian theory that continuous variation in a population is the source of evolutionary change. See: Darwinism, continuous evolution, evolution, Mendelism.

divine watchmaker: a version of the argument from design, advocated by William Paley. Paley argued that if one did not know what a watch was, one would still assume it to be designed, given its complexity; so too with living things. See: argument from design, natural theology.

dominance: a Mendelian term used to describe the phenomenon by which an offspring contains two different genes for a given characteristic but physically exhibits the characteristic for only one of the genes. The dominant gene masks the effects of the other gene, which is defined as recessive. See: Mendelism.

dorsal lip: a region of the blastula that functions as an organizer. See: blastula, organizer.

ELSI: an acronym for "ethical, legal, and social issues" that might arise from genome research. The United States and other countries typically organize and fund ELSI programs as part of their genome projects.

emboitement: the most popular preformation theory, espoused by Jan Swammerdam in 1672. See: preformationism.

embryonic induction: see: induction

embryonic stem cell: a stem cell in a developing embryo that contains the potential to differentiate into any cell type. See: stem cell.

endoplasmic reticulum: a cellular structure that is the site of protein production.

enzyme: a protein molecule capable of initiating and accelerating biochemical reactions.

epigenesis: an eighteenth-century theory stating that in the process of generation, life acquires form from unformed matter, through a process that is unique to living things. It is the opposite of preformationism. See: preformationism, generation.

epistasis: the phenomenon of multiple genes interacting to form a physical trait. See: additive genetics.

ether: a hypothetical substance once believed to fill the entire universe. Ether was invoked by mechanical philosophers to explain phenomena such as gravity or magnetism, in which objects seem to have an effect on each other even though they do not come into physical contact. Ether, it was believed, provides an indirect form of contact. See: mechanical philosophy.

eugenics: a term coined in 1883 by Francis Galton as "the study of the agencies under social control that may improve or impair the racial qualities of future generations, either physically or mentally." Eugenics became popular internationally in the early twentieth century as many countries initiated mandatory sterilization and anti-immigration laws against those deemed genetically unfit (generally, poor and/or nonwhite members of society).

evolution: the nineteenth-century theory, commonly associated with Jean-Baptiste de Lamarck and Charles Darwin, stating that living things can change their form over time.

expression: the activation of a gene sequence in order to produce a protein.

feeblemindedness: a category of mental illness, vaguely defined as encompassing a low level of mental, moral, and/or social fitness, that was accepted in the late nineteenth and early twentieth centuries.

fermentation: the breakdown of complex organic molecules into simpler molecules. Often refers to the breakdown by yeast of sugars into alcohol.

gastrula: an early-stage embryo that contains the inner cell mass. See: inner cell mass.

gel electrophoresis: a technique for separating molecules of different sizes using an electric current.

gene: a term coined in 1909 by Wilhelm Johannsen, describing the fundamental unit of hereditary material predicted by Mendelism; discovered in the early 1950s to be encoded by deoxyribonucleic acid (DNA). See deoxyribonucleic acid (DNA), Mendelism.

gene gun: a device that allows for the random insertion of genes into plants; DNA is used to coat tungsten particles, which are subsequently fired at high velocity into plant cells.

generation: a term used to describe both heredity and development; used prior to the twentieth century, when heredity and development were assumed to be the same process; became obsolete upon discovery of the continuity of the germplasm. See: continuity of the germplasm.

genetic code: a list of which DNA base triplets encode which amino acids. See: deoxyribonucleic acid (DNA), nucleotide, triplet.

genetic crossover: an exchange of chromosome material during meiosis.

genetic discrimination: discrimination against individuals or groups because of their actual or perceived genetic makeup.

genetic disease: traditionally, a disease that exhibits a pattern of Mendelian inheritance. Increasingly used to describe any disease that seems to have a genetic component. See: Mendelian inheritance.

genetic drift: a theory stating that random fluctuations of gene frequencies can have profound effects on the genetic makeup of a small interbreeding population and can therefore be an important component of evolutionary change.

genetic engineering: see: recombinant DNA technology, biotechnology.

genetic linkage: a term describing genes that are transmitted to offspring together more often than would be expected by chance because the genes are in close proximity on a chromosome and that therefore the probability of a genetic crossover between the genes is low. See: independent assortment, genetic crossover, genetic mapping.

genetic mapping: the use of techniques for determining genetic linkage to calculate relative distances between genes on a chromosome. See: gene, chromosome, genetic linkage.

genetic risk: see genetic susceptibility.

genetics: the study of the hereditary function of the gene. Coined in 1905 by William Bateson to replace the term *Mendelism*. See: Mendelism, gene.

genetic susceptibility: a term describing an increased risk for a given disease because of the presence of specific gene variants that are associated with that disease; also called genetic risk.

genome: the full set of genes in an organism. See: gene.

genomics: the study of genome structure and function. See: genome.

germplasm: a term coined by August Weismann to describe a set of cells that is responsible for passing on the hereditary material to offspring. See: continuity of the germplasm.

gravity: a force introduced in Isaac Newton's *Principia Mathematica* to account for the mutual attraction of two distant objects in space.

Great Chain of Being: an ancient theory of living things that states that all organisms can be arranged linearly in terms of a hierarchy of progressive complexity, culminating in humans.

hard inheritance: the term frequently used to describe and defend August Weismann's theory of the continuity of the germplasm; advocated by those who supported Weismann and opposed the theory of the inheritance of acquired characteristics. See: continuity of the germplasm, inheritance of acquired characteristics.

heritability: a measure of the proportion of variation in a population of organisms that is of genetic rather than environmental origin.

Human Genome Project: a coordinated, international effort to produce genetic and physical maps and sequence information for the complete genomes of humans and model organisms. See: genetic map, physical map, sequencing.

hybridization: the breeding of two pure-breeding lines of organisms to obtain hybrid offspring. See: pure-breeding.

independent assortment: the phenomena by which genes are transmitted independently to offspring, assuming that there is no linkage between genes. See: genetic linkage.

induction: the triggering of a developmental process in an embryo.

inherent tendency toward complexity: Jean-Baptiste de Lamarck's theory that evolution occurs by way of progressive increases in the complexity of individual organisms and their progeny, beginning with the spontaneous generation of simple organisms and, given enough time, culminating in humans.

Lamarck essentially applied evolution to the Great Chain of Being to form his theory. See: evolution, spontaneous generation, Great Chain of Being.

inheritance of acquired characteristics: a now-discredited theory of inheritance, commonly associated with Jean-Baptiste de Lamarck but in fact broadly accepted by natural philosophers, including Charles Darwin. The theory states that organisms can inherit physical characteristics that their parents have acquired during their lifetimes; for example, if a blacksmith, in his work, uses his right arm more than his left, his offspring may be born with a right arm stronger than the left.

initial public offering (IPO): the first sale of a corporation's public shares to investors.

inner cell mass: a set of pluripotent cells in a gastrula that subsequently differentiates into an embryo. See: pluripotent, gastrula.

intermediate forms: organisms with inherited characteristics intermediate between their parents. This concept was used by several natural philosophers, notably Carolus Linnaeus, to argue against evolution; they stated that new species are produced not by evolution but by the hybridization of two preexisting species to produce an intermediate form. See: evolution, hybridization.

in vitro: outside the body.

in vivo: inside the body.

lac operon: a system of interacting genes that together regulate the expression of lactose protein in bacteria. See: gene, expression.

linkage: see genetic linkage.

macromolecules: very large molecules, such as proteins and DNA.

mechanical philosophy: a theory of mechanism outlined by René Descartes in his 1637 *Discours de la Méthode* (Discourse on Method). See: mechanism.

mechanism: the theory that all phenomena, including life, are explainable with reference to simple mechanical principles. The opposite of vitalism. See: vitalism.

meiosis: see reduction division.

Mendelian inheritance: an expected, predictable pattern of inheritance of a given trait that is encoded by one or several genes.

Mendelism: named after Gregor Mendel, the theory that hereditary characters are transmitted by way of discrete units of hereditary information.

messenger RNA: an RNA molecule produced in the nucleus as a copy of DNA sequence, which migrates to the ribosomes and acts as a template for protein synthesis. See: ribonucleic acid (RNA), deoxyribonucleic acid (DNA), ribosome.

microsome: a molecule composed of ribonucleic acid and protein, located on the endoplasmic reticulum and responsible for producing proteins. See: endoplasmic reticulum.

mitosis: the process of cellular replication.

modern synthesis: a term used to describe mid-twentieth century research that reconciled Mendelism and Darwinism. See: Mendelism, Darwinism.

molecular biology: a term coined by Warren Weaver in 1938 to describe the study of the structure and function of biological molecules and subcellular structures, for the specific purpose of identifying and understanding the physical and chemical nature of the gene.

***molécules organiques* (organic molecules):** theoretical structures described by George-Louis LeClerc, Comte de Buffon, which were indivisible and unique to life and which were poured into an organism's *moules intérieurs* (interior mold) in order to give the organism form. Part of a theory of development seen by Buffon as a compromise between preformationism and epigenesis. See: epigenesis, preformationism, *moules intérieurs.*

Morgan: a unit of distance used to measure genetic linkage. One Morgan is divided into 100 centiMorgans (cM); one cM corresponds to a 1 percent probability of genetic crossover between two genes. See: genetic linkage, genetic crossover.

***moules intérieurs* (interior molds):** theoretical molds in an organism, described by George-Louis LeClerc, Comte de Buffon, in which his *molécules organiques* are poured. Part of a theory of development seen by Buffon as a compromise between preformationism and epigenesis. See: epigenesis, preformationism, *molécules organiques.*

multifactorial: see complex trait.

natural faculties: Galen of Pergamum's term for the vitalistic properties of internal organs. See: vitalism.

natural history: the nonexperimental study of nature, using methods such as observation, classification, description, and organization.

natural philosophy: A term used from the fourteenth to the nineteenth centuries for explanations of the natural world and its relationship to God. Natural philosophy was distinct from the Scholasticism in that it avoided (at least in its rhetoric) any reference to occult or religious explanations for natural phenomena. See: Scholasticism.

natural selection: Charles Darwin's 1859 theory of evolution, still widely accepted today, stating that those organisms with preexisting characteristics best adapted to their environments will tend to survive and procreate more often than organisms that are less well adapted. These "fitter" organisms will pass on their adaptive traits to their offspring. In time, this process will result in large-scale changes in organismal structure, or evolution. See: evolution.

natural theology: a late-seventeenth-century movement that described nature's complexity as evidence of God's perfection. See: argument from design.

naturphilosophie: a nineteenth-century German philosophical movement that saw all of nature as the expression of self-consciousness; expanded vitalism to the whole universe, arguing that all of nature was like a living organism guided by vital forces. See: vitalism.

neural tube: An early embryonic structure from which the brain and spinal cord are formed.

nucleotide: the fundamental structural unit of DNA and RNA, consisting of a sugar molecule, a phosphate molecule, and one of four bases: adenine, thymine (or uracil for RNA), guanine, or cytosine. See: deoxyribonucleic acid, ribonucleic acid.

nucleus: a cellular structure discovered in the late nineteenth century, eventually found to contain the chromosomes.

occult: For those who advocated mechanical philosophy, any explanation that referred to nonmechanical causes of phenomena.

operator: DNA sequence in the lac operon that, if bound by a repressor protein, prevents lactose expression. See: lac operon, expression, deoxyribonucleic acid (DNA), repressor.

organizer: embryonic tissue that initiates the process of embryonic development. See: induction.

Pangenesis: Charles Darwin's theory of inheritance and his explanation for variation in a population. Darwin argued that organisms can pass on acquired characteristics to their offspring in the form of gemmules, small bits of matter that migrate from all parts of the body to the reproductive organs.

parthenogenesis: a phenomenon seen in some organisms in which a female gives birth in the absence of fertilization by a male.

pedigree: a visual depiction of the transmission of an inherited trait within a biological family.

personalized medicine: a view that in the near future drugs will be tailored to the unique genetic constitution of an individual (or small sets of individuals), rather than administered to whole populations of patients.

phage: viruses; simple quasi-organismal structures consisting of nucleic acid surrounded by a protein coat.

pharmacogenetics: the study of genetic variation in both individual responsiveness to drugs and individual risks of side effects.

pharmacogenomics: whole-genome analysis of relationships between genetic variations and differential drug responses.

physical mapping: the determination of physical distances among DNA sequences along a chromosome. See: deoxyribonucleic acid (DNA), chromosome.

plasmid: a short, circular DNA strand that occurs naturally in bacteria and that is commonly used in recombinant DNA experiments. See: deoxyribonucleic acid (DNA), recombinant DNA technology.

pluripotent: an adjective describing the capacity of a stem cell to differentiate into some but not all cell types in an organism. See: stem cell, totipotent.

polymer: a large molecule containing repeating units of molecular structure.

polymerase chain reaction: a technique used to rapidly identify and exponentially replicate a segment of DNA with a known sequence out of a complex mixture of DNA.

polymorphisms: variations in DNA sequence in different individuals at the same DNA location.

population genetics: the statistical study of gene frequency change in populations of organisms, pioneered in the 1920s and 1930s by Ronald Fisher, Sewall Wright, and J.B.S. Haldane.

postgenomics: an umbrella term referring to research using the data output of the Human Genome Project.

preformationism: A mechanistic theory stating that all organisms have always existed fully formed within their parents. The theory was a means by which mechanists could argue that organisms did not acquire form during development, which to mechanists sounded vitalistic. In contrast to epigenesis. See: mechanism, vitalism, epigenesis.

probe: a small strand of radioactive DNA with a known base sequence. DNA probes can be bound to DNA with a base pair sequence that matches that of the probe, allowing the detection of very specific DNA sequences from a complex DNA mixture. See: deoxyribonucleic acid (DNA), nucleotide.

promoter: DNA sequence that, when activated, initiates expression of its associated gene. See: deoxyribonucleic acid (DNA), expression.

proteomics: the study of the proteome, the full protein complement of the genome. See: genome.

protoplasm: a term used in the late nineteenth century to describe the non-nuclear material in a cell. Now called cytoplasm. See: cell theory, nucleus.

public health genetics: the application of genetics to public health research and practice.

pure-breeds: organisms that always produce offspring with characteristics identical to their own.

recessive: see: dominance.

recombinant DNA technology: a collection of techniques that allow for experimental manipulation of DNA.

reduction division: now called meiosis; a form of cell division that results in progeny cells with half the normal chromosomal material; occurs during production of germ cells.

reductionism: explanations for studies of complex phenomena by reference to their simpler components. In biology, reductionism usually refers to explaining and/or studying biological processes using physics and/or chemistry.

repressor: a protein that, if bound to the operator in the lac operon, prevents lactose expression. See: lac operon, operator, expression.

restriction enzymes: a class of enzymes in bacteria that cut DNA at very specific locations by recognizing specific base sequences. See: enzyme, deoxyribonucleic acid (DNA).

restriction fragment length polymorphism (RFLP): DNA sequence that, when cut with a restriction enzyme, produces differently sized fragments in different individuals.

ribonucleic acid (RNA): a molecule similar to DNA but containing as a sugar ribose in place of deoxyribose, and as a base uracil in place of thymine.

ribosome: see microsome.

Scholasticism: a mixture of Christian theology and ancient Greek philosophy that was the chief domain of intellectual thought during the medieval period (fifth through sixteenth centuries).

semi-conservative replication: the process by which DNA replicates itself. Double-stranded DNA separates into two separate strands, each of which acts as a template for the construction of a new strand. The two resulting DNA molecules thus have one each of the old strand and the new strand of DNA.

sequencing: determining the exact order of bases in a DNA sequence. See: deoxyribonucleic acid (DNA).

shifting balance theory of evolution: a theory of evolution, first espoused by Sewall Wright in 1930, in which genetic drift rather than natural selection is the primary cause of evolutionary change; according to the theory, evolution is thus primarily a result of random, chance occurrences, rather than the natural selection of adaptive traits.

somatic cell hybridization: the production of hybrid cells that contain mouse and human chromosomes.

Southern Blot: a technique that uses gel electrophoresis and DNA probes to identify DNA with a specific base sequence. See: gel electrophoresis, probe, deoxyribonucleic acid (DNA).

spontaneous generation: an ancient theory stating that livings things could form from nonliving matter.

stem cell: a cell that has the potential to differentiate into many other cell types in the body. Stem cells function to replenish the body with specific cell types as needed.

sticky ends: Single-stranded base pairs at the site of a DNA sequence cut by a restriction enzyme. See: restriction enzyme, deoxyribonucleic acid (DNA), nucleotide.

swamping: a term used to describe what was seen as a problem in continuous inheritance theories, whereby new variation can never be maintained once introduced into a population of interbreeding organisms. See: continuous inheritance.

systems approach: the holistic study of complex systems, focusing on the interactions and relationships among different components of a system. In contrast to a reductionist approach, which involves isolating and studying individual components of a system.

teleology: literally, the study of the end; used to describe explanations that make reference to the purpose of an entity. For example, "A hand has an opposable thumb in order to grasp objects" is a teleological explanation.

totipotent: an adjective describing the capacity of a stem cell to differentiate into any type of cell. See: stem cell, pluripotent.

transfer RNA: RNA molecules that carry amino acids to the ribosome for protein synthesis. See: ribonucleic acid (RNA), amino acid, ribosome.

triplet: Three DNA base pairs that, together, encode one amino acid. See: genetic code, nucleotide, deoxyribonucleic acid.

vitalism: the theory that there exists a substance or property that is unique to and essential for living things to be alive and that is not reducible to the normal laws of physics or chemistry. The opposite of mechanism. See: mechanism.

x-ray diffraction: the use of x-rays to study molecular structure; a molecule is bombarded by x-rays, and the pattern of dispersion of the x-ray particles—the x-ray *diffraction* pattern—provides information about the molecule's structure.

yeast artificial chromosome (YAC): a laboratory-constructed chromosome that can be inserted into yeast. Used for inserting large amounts of foreign DNA into yeast cells. See: chromosome.

BIBLIOGRAPHY

Advisory Committee on Health Research. 2002. *Genomics and World Health.* Geneva: World Health Organization.

Allen, Garland. 1978. *Thomas Hunt Morgan: The Man and His Science.* Princeton: Princeton University Press.

Bacon, Francis. 1629 *Instauratio Magna* [The Great Instauration]. The Constitution Society, http://www.constitution.org/bacon/instauration.htm.

Bateson, William. 1894. *Materials for the Study of Variation.* London: Macmillan.

Bateson, William. 1902. *Mendel's Principles of Heredity: A Defence.* London: Cambridge University Press.

Bowler, Peter. 1984. *Evolution: The History of an Idea.* Berkeley: University of California Press.
    A broad and useful introductory survey of the history of evolutionary theory.

Bowler, Peter. 1989. *The Mendelian Revolution: The Emergence of Hereditarian Concepts in Modern Science and Society.* Baltimore: Johns Hopkins University Press.
    A history of Mendelism, emphasizing the differences between Mendel's research and the "Mendelism" that subsequently came to prominence after Mendel's rediscovery.

Bud, Robert. 1993. *The Uses of Life: A History of Biotechnology.* Cambridge: Cambridge University Press.
    A detailed history of biotechnology, encompassing its prehistorical roots in early fermentation through to post-recombinant DNA biotechnology.

Buffon, Comte de. 1749. *Histoire Naturelle* [Natural History]. New Jersey City University, http://faculty.njcu.edu/fmoran/buffonhome.htm.

Buffon, Comte de. 1971 [1779]. *Des Epoques de la Nature* [Epochs of Nature]. Paris: Éditions Rationalistes.

Burke, Wylie, Muin J. Khoury, and Elizabeth J. Thomson, eds. 2000. *Genetics and Public Health in the 21st Century: Using Genetic Information to Improve Health and Prevent Disease.* Oxford: Oxford University Press.

Charles, Daniel. 2001. *Lords of the Harvest: Biotech, Big Money, and the Future of Food.* Cambridge, MA: Perseus.

A balanced, entertaining, and detailed history of the rise of the genetically modified foods industry.

Coghlan, Andy, and Nell Boyce. 2006. "Human Genome: The End of the Beginning." *New Scientist,* http://www.newscientist.com/article/mg19225780.050-human-genome-the-end-of-the-beginning.html.

Coleman, William. 1965. "Cell, Nucleus and Inheritance: An Historical Study." *Proceedings of the American Philosophical Society* 109: 124–158.

A history of studies of the cell with a focus on its hereditary functions, moving from Darwin's Pangenesis to the discovery of the chromosomes and the nucleus.

Conrad, Peter, and Jonathan Gabe, eds. 1999. *Sociological Perspectives on the New Genetics.* Oxford: Blackwell.

A set of essays addressing a broad diversity of issues related to both the meaning and the impact of modern genetics.

Cook-Deegan, Robert. 1996. *The Gene Wars: Science, Politics and the Human Genome.* New York: Norton.

An excellent history of the Human Genome Project from the perspective of an insider analyst at the Office of Technology Assessment, which played a considerable role in early policy efforts to form an American project.

Darlington, Cyril. 1932. *Recent Advances in Cytology.* London: Churchill.

Darlington, Cyril. 1939. *The Evolution of Genetic Systems.* Cambridge: Cambridge University Press.

Darwin, Charles. 1859. *The Origin of Species.* The Online Literature Library, http://www.literature.org/authors/darwin-charles/the-origin-of-species/index.html.

Darwin, Erasmus. 1794. *Zoonomia.* Project Gutenberg, http://www.gutenberg.org/etext/15707.

Degler, Carl N. 1992. *In Search of Human Nature.* Oxford: Oxford University Press.

A history of twentieth-century social sciences in the United States and their relationship to the growth in prominence of Darwinism.

Descartes, René. 1637. *Discours de la Méthode* [Discourse on Method]. The History of Computing Project, http://www.thocp.net/biographies/papers/discours_dela_methode.htm.

Dobzhansky, Theodosius. 1937. *Genetics and the Origin of Species.* National Academies Press, http://www.nap.edu/catalog/5923.html.

Fisher, Ronald A. 1918. "The Correlation between Relatives on the Supposition of Mendelian Inheritance." *Philosophical Transactions of the Royal Society of Edinburgh* 52: 399–433.

Fisher, Ronald A. 1922. "On the Dominance Ratio." *Proceedings of the Royal Society of Edinburgh* 42: 321–341.

Fisher, Ronald A. 1928. "The Possible Modification of the Response of the Wild Type to Recurrent Mutations." *American Naturalist* 62: 115–126.

Fisher, Ronald A. 1930. *The Genetical Theory of Natural Selection.* Oxford: Clarendon Press.

Fleming, Donald. 1969. "Emigré Physicists and the Biological Revolution." In *The Intellectual Migration: Europe and America, 1930–1960,* ed. Donald Fleming and Bernard Bailyn (Cambridge, MA: Harvard University Press, 1969).

An excellent historical study of the movement of physicists into biology and their focus on reducing biological processes to physical laws.

Galen. 170 A.D. *On the Natural Faculties.* The Internet Classics Archive, classics.mit.edu/Galen/natfac.html.

Galen. 1992 [ca. 180 A.D.]. *De Semine* [On Seed]. Edited by Phillip De Lacy. Berlin: Akademie Verlag.

Gilbert, Scott. 1982. "Intellectual Traditions in the Life Sciences: Molecular Biology and Biochemistry." *Perspectives in Biology and Medicine* 26: 151–162.

A good, accessible introduction to the intertwining histories of biochemistry and molecular biology.

Gould, Stephen Jay. 1996. *The Mismeasure of Man.* New York: Norton.

A critical history of the use of biology to justify social inequality.

Grene, Marjorie. 1983. *Dimensions of Darwinism: Themes and Counter Themes in Twentieth Century Evolutionary Theory.* Cambridge: Cambridge University Press.

A collection of essays discussing various aspects of the decline and rise of Darwinism in the twentieth century.

Grew, Nehemiah. 1965 [1682]. *The Anatomy of Plants.* New York: Johnson Reprint Corporation.

Haldane, J.B.S. 1924. "A Mathematical Theory of Natural and Artificial Selection." Blackwell, http://www.blackwellpublishing.com/ridley/classictexts/haldane1.asp.

Haldane, J.B.S. 1990 [1932]. *The Causes of Evolution.* Princeton: Princeton University Press.

Hall, Stephen. 2002. *Invisible Frontiers: The Race to Synthesize a Human Gene.* 2nd ed. New York: Oxford University Press.

A wonderfully entertaining, comprehensive, and informative account of the race by three research groups to use recombinant DNA technology to clone insulin, set in the contexts of the rise of biotechnology and the recombinant DNA controversy.

Hamburger, Viktor. 1988. *The Heritage of Experimental Embryology: Hans Spemann and the Organizer.* Oxford: Oxford University Press.

Harvey, William. 1628. *De Motu Cordis* [On the Motion of Blood]. Modern History Sourcebook, http://www.fordham.edu/halsall/mod/1628harvey-blood.html.

Harvey, William. 1955 [1651]. *De Generatione* [On Generation]. Chicago: Encyclopedia Britannica.

Heilbron, John L., ed. 2003. *The Oxford Companion to the History of Modern Science.* New York: Oxford University Press.

An excellent collection of articles dealing with the history of science, organized thematically.

Hooke, Robert. 1665. *Micrographia.* Project Gutenberg, http://www.gutenberg.org/etext/15491.

Huxley, Julian. 1942. *Evolution: The Modern Synthesis.* New York and London: Harper and Brothers.

Huxley, Thomas Henry. 1868. "On the Physical Basis of Life." In Thomas Henry Huxley, *Autobiography and Selected Essays.* Project Gutenberg, http://www.gutenberg.org/etext/1315.

International Human Genome Consortium. 2001. "Initial Sequencing and Analysis of the Human Genome." *Nature* 409: 860–921.

The official first draft of the human genome sequence published by the public Human Genome Project.

Joint Centre for Bioethics. 2002. *Top Ten Biotechnologies for Improving Health in Developing Countries.* Toronto: University of Toronto Joint Centre for Bioethics.

Kay, Lily. 1993. *The Molecular Vision of Life: Caltech, the Rockefeller Foundation, and the Rise of the New Biology.* Oxford: Oxford University Press.

A detailed history of the early years of molecular biology, focused on the Rockefeller Foundation's role, its interests in biology-mediated social reform, and its funding of the Division of Biology at Caltech.

Kay, Lily. 2000. *Who Wrote the Book of Life? A History of the Genetic Code.* Stanford: Stanford University Press.

A history of the search for the genetic code, situated nicely in the context of Cold War interests in cryptanalysis and the increasing interpretation of the gene as encoded information.

Kevles, Daniel J. 1985. *In the Name of Eugenics: Genetics and the Uses of Human Heredity.* Cambridge, MA: Harvard University Press.

A comprehensive history of the eugenics movement, focused primarily on the United States.

Kevles, Daniel, and Leroy Hood. 1992. *The Code of Codes: Scientific and Social Issues in the Human Genome Project.* Cambridge, MA: Harvard University Press.

A excellent collection of essays describing the science, technology, and social context of the Human Genome Project.

Kimmelman, Barbara. 1983. "The American Breeders' Association: Genetics and Eugenics in an Agricultural Context, 1903–1913." *Social Studies of Science* 13: 163–204.

A history of how Mendelism and eugenics were related to and encouraged by the growth of agriculture in the United States.

Krimsky, Sheldon. 1982. *Genetic Alchemy: The Social History of the Recombinant DNA Controversy.* Cambridge, MA: MIT Press.

de Lamarck, Jean-Baptiste. 1809. *Philosophie Zoologique* [Zoological Philosophy]. Ian Johnston, Malaspina University-College, http://www.mala.bc.ca/~johnstoi/lamarck/tofc.htm.

Lindberg, David. 1992. *The Beginnings of Western Science.* Chicago: University of Chicago Press.

An excellent introduction to the history of science in the ancient and medieval worlds, written clearly and concisely and incorporating explanations of science, technology, and social context.

MacKenzie, Donald. 1981. *Statistics in Britain, 1865–1930.* Edinburgh: Edinburgh University Press.

A history of how the development of statistics under Francis Galton and Karl Pearson was intimately connected to their eugenic goals.

MacKenzie, Donald, and Barry Barnes. 1979. "Scientific Judgment: The Biometry-Mendelian Controversy." In *Natural Order: Historical Studies of Scientific Culture,* ed. Barry Barnes and Steven Shapin. Beverly Hills, CA: Sage, pp. 191–210.

A history of the debate between the Darwinists and the Mendelists.

Magner, Lois. 1994. *A History of the Life Sciences.* 2nd ed. New York: Marcel Dekker.

A broad history of the biological sciences from the ancient Greeks to the late twentieth century.

Maienschein, Jane. 1985. "History of Biology." *Osiris* 1: 147–162.

A survey of some of the major works in the history of the twentieth-century biological sciences in the United States.

Malebranche, Nicolas. 1980 [1674]. *De la Recherche de la Vérité* [On the Search for Truth], trans. Thomas M. Lennon and Paul J. Olscamp. Columbus: Ohio State University Press.

Mayr, Ernst. 1942. *Systematics and the Origin of Species.* New York: Columbia University Press.

Mayr, Ernst, and William Provine, eds. 1980. *The Evolutionary Synthesis.* Cambridge, MA: Harvard University Press.

A detailed history of the modern synthesis.

Mendel, Gregor. 1865. "Versuche über Pflanzen-Hybriden" [Experiments on Plant Hybrids]. MendelWeb, http://www.mendelweb.org/Mendel.html.

Morange, Michel. 1998. *A History of Molecular Biology*. Cambridge, MA: Harvard University Press.

A lucid and readable history of molecular biology from the 1930s to the end of the twentieth century.

Morgan, Thomas Hunt. 1934. *Embryology and Genetics*. New York: Columbia University Press.

Morgan, Thomas Hunt, Alfred Sturtevant, Hermann Muller, and Calvin Bridges. 1915. *The Mechanism of Mendelian Heredity*. Electronic Scholarly Publishing, http://www.esp.org/books/morgan/mechanism/facsimile/title3.html.

National Institutes of Health. 2006. "NIH Stem Cell Information Home Page." U.S. Department of Health and Human Services, http://stemcells.nih.gov/index.

An excellent introductory resource for more information about the science and current U.S. policy regarding stem cell research.

Nelkin, Dorothy, and M. Susan Lindee. 1995. *The DNA Mystique: The Gene as a Cultural Icon*. New York: Freeman.

A survey of the popularity of genetics in society and the consequent growth of exaggerated views about the power of DNA.

Newton, Isaac. 1972 [1726]. *Philosophiae Naturalis Principia Mathematica* [Mathematical Principles of Natural Philosophy]. 3rd ed. Cambridge, MA: Harvard University Press.

Office of Technology Assessment. 1988. *Mapping Our Genes: How Big? How Fast?* Washington, DC: U.S. Congress.

Olby, Robert. 1979. "Mendel No Mendelian?" *History of Science* 17: 53–72.

Olby argues that Mendel was not consciously seeking or describing the laws of genetics but instead was testing the theory of speciation by way of intermediate forms.

Paley, William. 1802. *Natural Theology*. University of Michigan Humanities Text Initiative, http://www.hti.umich.edu/cgi/p/pd-modeng/pd-modeng-idx?type=header&id=PaleyNatur.

Provine, William. 1971. *Origins of Theoretical Population Genetics*. Chicago: University of Chicago Press.

A detailed history of the work of Ronald Fisher, Sewall Wright, and J.B.S. Haldane.

Ray, John. 1691. *The Wisdom of God Manifested in the Works of the Creation*. The John Ray Initiative, http://www.jri.org.uk/ray/wisdom/index.htm.

Rheinberger, Hans Jorg. 2004. Gene. Stanford Encyclopedia of Philosophy, http://plato.stanford.edu/entries/gene/.

Rose, Geoffrey. 1992. *The Strategy of Preventive Medicine*. Oxford: Oxford University Press.

Sayre, Anne. 1975. *Rosalind Franklin and DNA*. New York: Norton.
    An excellent biography of Rosalind Franklin's work in elucidating the structure of DNA, set in the context of the difficulties faced by women scientists in mid-twentieth-century Britain.

Schrödinger, Erwin. 1944. *What Is Life?* Cambridge: Cambridge University Press.

Simpson, George Gaylord. 1944. *Tempo and Mode in Evolution*. New York: Columbia University Press.

Stevens, Tim. 2000, December 11. "Technology Leaders of the Year." *Industry Week* (online).

Swammerdam, Jan. 1672. *Miraculum Naturae* [The Miracle of Nature]. National Library of France, http://gallica.bnf.fr/ark:/12148/bpt6k98784j. capture.

Thackray, Arnold, ed. 1998. *Private Science: Biotechnology and the Rise of the Molecular Sciences*. Philadelphia: University of Pennsylvania Press.
    A set of essays that explore the meaning of public and private science in the context of molecular biology and biotechnology.

U.S. Department of Energy. 2006. "Ethical, Legal, and Social Issues." Department of Energy, http://www.ornl.gov/sci/techresources/Human_Genome/elsi/elsi.shtml.
    An excellent summary of various issues studied under the Human Genome Project's "Ethical, Legal, and Social Issues" (ELSI) program.

Various Authors. 2005. "Commentary: Systems Biology." *Cell* 121: 503–513.
    A collection of short scientific review articles discussing the meaning and progress of systems biology in the context of postgenomics.

Venter, J. Craig, Mark Adams, Eugene Myers, Peter Li, et al. 2001. "The Sequence of the Human Genome." *Science* 291: 1304–1351.
    Celera Genomics' announcement of the draft sequence of the human genome, published simultaneously with the public project's announcement.

Virchow, Rudolf. 1860 [1858]. *Die Cellularpathologie* [Cellular Pathology]. London: Churchill Books.

de Vries, Hugo. 1889. *Intracellular Pangenesis*. Electronic Scholarly Publishing, http://www.esp.org/books/devries/pangenesis/facsimile/.

de Vries, Hugo. 1909 [1901] *Die Mutationstheorie* [The Mutation Theory]. Chicago: Open Court Press.

Watson, James. 1968. *The Double Helix*. New York: Simon and Schuster.
    An entertaining, idiosyncratic, contextualized, and subjective history of Watson and Crick's discovery of the structure of DNA.

Watson, James, and Francis Crick. 1953. "A Structure for Deoxyribose Nucleic Acid." *Nature* 171: 737–738. Nature Publishers, www.nature.com/nature/dna50/watsoncrick.pdf.

Weismann, August. 1883. *Uber die Vererbung* [On Heredity]. Jena: Fischer.

Weismann, August. 1892. *Das Keimplasm* [The Germplasm]. Jena: Fischer.

White, Michael. 1945. *Animal Cytology and Evolution.* Cambridge: Cambridge University Press.

Wolff, Caspar Friedrich. 1896 [1759]. *Theoria Generationis* [Theory of Generation]. Leipzig: Engelmann Press.

Wright, Sewall. 1930. *"The Genetical Theory of Natural Selection:* A Review." *Journal of Heredity* 21: 340–356.

Wright, Susan. 1986. "Recombinant DNA Technology and Its Social Transformation, 1972–1982." *Osiris* 2: 303–360.

    A history of recombinant DNA technology, focused on the connections between science and technology, industrial and political interests, and public responses to the technology.

Wright, Susan. 1994. *Molecular Politics: Developing American and British Policy for Genetic Engineering.* Chicago: University of Chicago Press.

    A political and social comparative history of post-recombinant DNA regulation in Britain and the United States.

INDEX

About the Author

TED EVERSON is the Program Manager for Biotechnology Studies in the Center for Contemporary History and Policy at the Chemical Heritage Foundation. He earned a PhD in History and Philosophy of Science and Technology from the University of Toronto, in which he explored historically the increasing use of genetic concepts and technologies in healthcare, an MS in Medical Genetics from the University of British Columbia, focusing on genetic and physical mapping of the human genome, and a BS in Biology from the University of British Columbia. In addition to several academic articles and presentations for various audiences, he is the author of "Genetic Engineering: Methods," in Colin Hempstead and William Worthington, eds., *Encyclopedia of Twentieth Century Technology* and "Genetics and Molecular Biology," in Brian Baigre, ed., *History of the Exact Sciences and Mathematics*.